Energy Utilisation: The Opportunities and Limits

Wolfgang Osterhage

Energy Utilisation: The Opportunities and Limits

Physical Fundamentals and Technologies

 Springer

Wolfgang Osterhage
Wachtberg-Niederbachem, Germany

ISBN 978-3-030-79406-4   ISBN 978-3-030-79404-0   (eBook)
https://doi.org/10.1007/978-3-030-79404-0

This Springer imprint is published by the registered company Springer Nature Switzerland AG
The registered company address is: Gewerbestrasse 11, 6330 Cham, Switzerland

Preface

The talk about energy transition is prevailing everywhere and is occupying the media since 2011. The same goes for "renewable" energies. Against this backdrop, this book wants to introduce the reader to a compressed overview of some of the most important foundations of physics. At the same time, it is shown that any type of energy allocation or transformation can be traced back to atomic or nuclear causes. Furthermore, the current and well-known energy utilization technologies are presented. Treated in succession are classical steam power plants, nuclear energy, solar energy, wind power, bio mass and bio gas, geothermal heat and hydrodynamic as well as tidal power, rain power plants and combined heat and power generation. In addition, storage technologies are discussed. Finally, a perspective is given with respect to fusion and fuel cells. The appendix gives a comprisal of the current discussion concerning climate engineering in connection with the consequences of energy utilization technologies.

A German version of the book has been published by Springer in 2019, based on a series of lectures I held in the 2011/12 winter term at the Goethe University in Frankfurt. The present book has been re-worked and enhanced.

I would like to thank Anthony Doyle and Rajan Muthu for their constructive support of this project.

Wachtberg-Niederbachem,
Germany

Wolfgang Osterhage

Contents

1	**Introduction**	1
2	**Energy Balances**	5
	2.1 Energy Supply	5
	2.2 Fundamentals of Thermodynamics	8
	2.2.1 Introduction	8
	2.2.2 Energy	8
	2.2.3 Temperature	10
	2.2.4 First Law of Thermodynamics	13
	2.2.5 Second Law of Thermodynamics	16
	2.2.6 Heat Conduction	18
	2.2.7 Exergy and Anergy	19
	2.2.8 Enthalpy	26
	2.2.9 Energy Balances in General	28
3	**Physical Fundamentals of Energy Utilization**	31
	3.1 Electromagnetism	31
	3.1.1 Introduction	31
	3.1.2 Charge	31
	3.1.3 Current and Tension	33
	3.1.4 Magnetism	38
	3.1.5 Alternating Current	40
	3.1.6 Maxwell's Equations	42
	3.1.7 The Distribution of Electrical Energy	44
	3.2 Fluid Mechanics	46
	3.2.1 Introduction	46
	3.2.2 Liquids	46
	3.3 Gas Dynamics	52
	3.3.1 Ideal Gas	52
	3.3.2 Turbines	54
	3.4 Atomic Physics	57
	3.4.1 Introduction	57
	3.4.2 Radiation	58

3.4.3   Kirchhoff's Radiation Law 60
3.4.4   Planck's Radiation Law 61
3.4.5   Light Measurement 62
3.4.6   The Photo Effect 64
3.4.7   Photovoltaic Cell 68
3.4.8   The Compton Effect 71
3.4.9   Material Waves 73
3.4.10   Quantum Numbers 78
3.5   Nuclear Physics 81
3.5.1   The Periodic Table of the Elements 81
3.5.2   The Strong Interaction 82
3.5.3   Structure of the Atomic Nucleus 83
3.5.4   Nuclear Models 84
3.5.5   Radioactivity 87
3.5.6   Reactions in Nuclear Physics 99

4   Types of Energy Generation 103
4.1   Steam Power Plants 103
4.1.1   Introduction 103
4.1.2   Coal 104
4.1.3   Gas 105
4.1.4   Oil 106
4.1.5   Thermal and Combustion Power Plants: Origin
         of Fuels 106
4.1.6   Nuclear Power 107
4.2   Solar Power Plants 109
4.2.1   Parabolic Trough Power Plants 110
4.2.2   Photovoltaic 110
4.2.3   Solar Energy: Origin 111
4.3   Wind Power 112
4.3.1   Introduction 112
4.3.2   Realization 113
4.3.3   Wind Energy: Origin 115
4.4   Bio Mass 116
4.4.1   Combustion Technology 116
4.4.2   The Combustion Process 118
4.4.3   Fuel 119
4.4.4   The Generation of Electricity and Heat 120
4.4.5   Bio Mass: Origin 120
4.5   Bio Gas 121
4.5.1   Composition and Characteristics of Bio Gas 122
4.5.2   Technological Pre-conditions 123
4.6   Geothermal Heat 124
4.6.1   Introduction 124
4.6.2   Heat Pump Systems 125

	4.6.3	Geothermal Power Plants	125
	4.6.4	Geothermic Energy: Origin	125
4.7	Hydrodynamic Power		126
	4.7.1	Introduction	126
	4.7.2	Barrages	126
	4.7.3	Pumped Storage Power Plants	127
	4.7.4	Hydrodynamic Power: Origin	128
4.8	Rain Power Stations		128
	4.8.1	Background	128
	4.8.2	Concept	128
4.9	Combined Heat and Power Generation		129
4.10	Tidal Power Stations		130

5 Storage Technologies 133
5.1	Transmission Losses		134
	5.1.1	Voltage Levels	134
	5.1.2	Conduction Losses	134
5.2	Storage		134
	5.2.1	Lead-Acid Batteries	135
	5.2.2	Lithium-Ion Batteries	135
	5.2.3	Lithium-Sulphur Batteries	135
	5.2.4	Lithium-Air Batteries	136
	5.2.5	Sodium High-Temperature Batteries	136
	5.2.6	Redox-Flow Batteries	136

6 Future Approaches 139
6.1	Nuclear Fusion		139
	6.1.1	Challenges	140
	6.1.2	Technical Questions	141
	6.1.3	The Stellarator	142
6.2	Fuel Cell		142

7 Smart Energy 145
7.1	Introduction	145
7.2	The Smart Energy Vision	145
7.3	Smart Meters	146
7.4	Smart Grid	147

8 Appendix: Climate Engineering 149
8.1	Point of Departure		149
8.2	The Basic Problem		150
8.3	Selective Interventions in the Climate System		150
	8.3.1	Causal Removal Technologies	151
	8.3.2	Technologies for Symptomatic Compensation of Climate Change	151
	8.3.3	State of Discussion	152
	8.3.4	Legal Context	153

	8.3.5	Costs	154
	8.3.6	Approaches	155
8.4	Detailed Technological Measures		155
	8.4.1	Starting Point	155
	8.4.2	Possibilities of Influence	156
	8.4.3	State of the Art	161
	8.4.4	Side Effects	161
8.5	Physical Background Considerations		162
	8.5.1	Energy Balances	162
	8.5.2	Chaos	162
	8.5.3	Fine Adjustment	163
8.6	Consolidation		164
	8.6.1	Irreversibility	164
	8.6.2	Arguments	164
	8.6.3	Risk Analysis	165
8.7	Reference Frame		165
8.8	Ethics		166
8.9	Conclusions		167

Problems and Solutions 169

Index 171

Chapter 1
Introduction

When transforming different types of energy, practically all disciplines of physics with the exception of high energy physics and cosmology play a role. In this book, the most important ones have been selected. These areas will be looked at in general. The following special subjects are concerned:

- thermodynamics and heat transfer,
- fluid dynamics and gas dynamics,
- electromagnetism,
- atomic physics,
- nuclear physics.

When, for example, considering a thermal power plant, the underlying schema in Fig. 1.1 is applicable.

The process depicted consists of the following components:

- combustion of coal, gas, oil or organic substances,
- a chain reaction in a nuclear reactor,
- air movement (wind),
- absorption of solar energy,
- absorption of geothermal heat,
- fermentation of organic substances.

Heat exchangers are capable to absorb heat generated directly or transfer it to a secondary circuit.

The mechanical converter is generally a turbine, the electrical one a generator.

There are processes that can do without one or another component. To these belong wind power plants (no heat exchanger) and photovoltaic assemblies (no heat exchanger, no mechanical converter, no electrical converter).

Before going into the details, we should start by looking at the finite energy supply at our disposal. After this, we will consider the first and second laws of thermodynamics as well as energy balances in depth, succeeded by the most important aspects

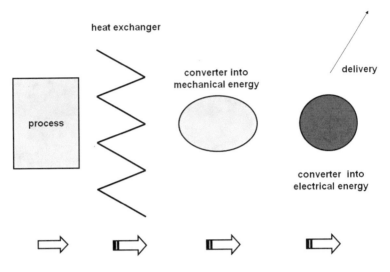

Fig. 1.1 Components of energy transformation

of fluid dynamics and electromagnetism, both playing important roles in energy utilization.

Besides this, there are fundamentals from which all types of energy can be derived, i.e. atomic and nuclear processes. These will be looked at separately in the relevant sections. To understand these connections we will give an overview of two more special areas of physics:

- atomic physics,
- nuclear physics.

Basically, we can retain the heat we obtain from two resources in nature:

- the sun's radiation and
- geothermal heat.

The sun's radiation in turn is generated by nuclear processes in the interior of the sun—nuclear fusion. The sun light itself is created by atomic processes by excitation and de-excitation in the electron shell of the sun's atoms. Geothermal heat is the result of the radioactive decay of unstable elements in the interior of the earth or the earth's crust.

Fossil fuels have resulted from the transformation of initially organic matter, which can be traced back to another atomic process, photo synthesis, or—regarding its animal type—via the food chain to plants. The same holds for bio mass. Wind power and hydrodynamic power are the results of mass flows from climatic effects and thus in the end also from the sun's radiation. Remains nuclear energy, which cannot be reduced any further to other natural processes.

Figure 1.2 gives an overview of these linkages.

Fig. 1.2 Linkage of energy types

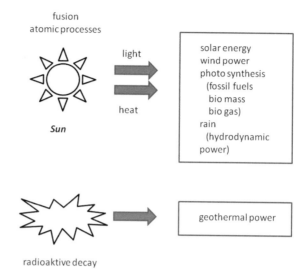

The present discussion around energy has discovered a connection between types of energy utilization and their influence on the earth's climate. At the end of the book, I have added an appendix dealing with the problems of climate engineering.

Chapter 2
Energy Balances

Abstract It is general knowledge in physics: energy is not renewable. Energy is never created nor renewed but at most transformed—and this by diminishing its useable component. We will start off with the term "force" and approach the definition of "energy" via the term "work". The basis of potential and kinetic energy will then allow us the further treatment of thermodynamics. Thus we will deal with motion, i.e. dynamics, and its triggering momentums, and with the motion of heat itself as well. After basic definitions concerning the quantity "temperature" we will get acquainted with the two fundamental laws of thermodynamics. They are decisive for our understanding, in which direction the world is moving and which are the scientific and technical possibilities at our disposal, and which are not.

2.1 Energy Supply

It is general knowledge in physics: energy is not renewable. Energy is never created nor renewed but at most transformed—and this by diminishing its useable component. This is true for the whole cosmos, for our living environment here on earth and for every technical assembly for power transformation. Let us consider the following question:

> What is the total amount of energy at our disposal after the Big Bang?

These are the most important basic cosmological data (Fig. 2.1).

The "Standard Hot Big Bang Model" is the cosmological model widely accepted today and is based on the fact that gravitation dominates the whole development of the universe, but that the observed details are determined by the laws of thermodynamics, hydrodynamics, atomic physics, nuclear physics and high energy physics. The graph in Fig. 2.2 illustrates the origin and development of our universe.

It is assumed that during the first second after the beginning temperature was so high that a complete thermodynamic equilibrium existed between photons, neutrinos,

Radius of Maximum Expansion	$18{,}94 \times 10^9$ LY
Time until Maximum	$29{,}76 \times 10^9$ Y
Age	13×10^9 Y
Present Density	$14{,}8 \times 10^{-30}$ g/cm^3
Present Volume	$38{,}3 \times 10^{84}$ cm^3
Density at Maximum	5×10^{-30} g/cm^3
Maximum Volume	114×10^{84} cm^3
Total Mass	$5{,}68 \times 10^{56}$ g
Number of Baryons	$3{,}39 \times 10^{80}$

Fig. 2.1 Basic cosmological data

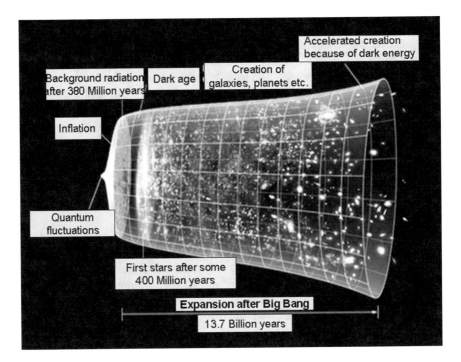

Fig. 2.2 The big bang model

electrons, positrons, neutrons, protons and different hyperons and mesons as well as possible gravitons.

After several seconds, the temperature dropped to about 10^{10} K, and the density amounted to about 10^5 g/cm^3. Particles and anti-particles had annihilated themselves, hyperons and mesons had decayed and neutrinos and gravitons had decoupled from matter. The universe now existed of free neutrinos and perhaps gravitons, the field quanta of gravitational waves.

In the following phase between 2 and about 1000 s, a first primordial creation of elements took place. Before, such attempts had been destroyed by high-energy protons. These elements consisted mainly of α-particles (He^4), traces of deuterium,

He^3 and Li, and accounted for 25%, the remainder was hydrogen nuclei (protons). All heavier elements were created later.

Between 1000 s and 10^5 years thereafter the thermal equilibrium was maintained by a continuous transfer of radiation into matter as well as permanent ionization processes and the building of atoms. In the end, temperature dropped to a few thousand degrees. Now the universe became dominated by matter instead of radiation. Photons were no longer sufficiently energetic to ionize hydrogen atoms permanently for example.

After photon pressure had disappeared condensation of matter into stars and galaxies could commence: between 10^8 and 10^9 years afterwards.

As can be seen from the basic data mass and energy of the universe are finite—even when taking the famous Einstein equation $E = mc^2$ into account we can only dispose of a limited energy supply. And from this again only a portion can be transformed into useable energy. During transformation—independently of its type—energy is always devalued such that the portion of useable energy diminishes constantly.

Now there could be an objection along the lines that cosmological models make use of yet unknown quantities like Dark Matter and Dark Energy, enlarging the total supply significantly with also unknown characteristics. The space telescope "Planck" has measured and mapped the Cosmic Background Radiation for four years. This led to a revision of earlier indications about Dark Matter. These indications were based on a surplus of positrons in cosmic radiation. One explanation could have been that this surplus would have been the result of collisions of particles with Dark Matter. This would, however, only have been plausible, if today the collision probability would be greater than at the time shortly after the Big Bang. But the Planck results disprove this hypothesis.

A different assumption is based on the theory that Dark Matter consists of WIMPs (Weakly Interacting Massive Particles). There is a wide consensus about the approximate total distribution of Dark Matter in the cosmos. Thus, the share of Dark Matter is larger in regions of greater mass accumulation like our Milky Way than elsewhere.

WIMPS do not react with baryonic mass particles the way we know it—like, for example, electrons with the shell of atoms. But scattering reactions with atomic nuclei can be imagined. The proof would then be the detection of an atomic nucleus having received such a kick. Two experiments to deliver this proof are being carried out. They are located in an underground laboratory in the Gran Sasso: XENON and CREST. The element xenon is used in many detectors as a scintillator. A xenon nucleus hit by a WIMP with hundred times the mass of a proton would deliver a characteristic scintillation track in a tank filled with liquid xenon. People have calculated that in 1 kilogram xenon such a collision could take place about every ten years. This means long experimental times. Of course, an upgrade of the xenon detector currently containing 3500 kg of liquid is already in planning.

The detector CRESST had been conceived for the search for lighter WIMPs. In this case, crystals are cooled down to shortly above absolute zero. After a collision with a WIMP, the temperature rises minimally; thus the event can be registered.

While Dark Matter slows down the expansion of the Universe by gravitation Dark Energy produces the opposite effect. Current theories are based on the dominance of

Dark Energy leading to an ever-expanding Universe. To achieve this, Dark Energy must have a negative pressure leading to repelling gravitation. Dark Energy itself is not capable of creating structures but can influence the structure of matter. It is expected that the study of certain supernova explosions gives indications about the influence of Dark Energy.

Independent of the amount of the dark components ever to be identified: as of today we do not dispose of them neither do we have an apparatus for their conversion. And even, if some day we would—even Dark Matter and Dark Energy will be finite as well.

2.2 Fundamentals of Thermodynamics

2.2.1 Introduction

We will start with the term "force" and approach the definition of "energy" via the term "work". The basis of potential and kinetic energy will then allow us the further treatment of thermodynamics. Thus we will deal with motion, i.e. dynamics, and its triggering momentums, and with the motion of heat itself as well.

After basic definitions concerning the quantity "temperature", we will get acquainted with the two fundamental laws of thermodynamics. They are decisive for our understanding, in which direction the world is moving and which are the scientific and technical possibilities at our disposal, and which are not.

2.2.2 Energy

Before entering into thermodynamics, we need to introduce the term "energy" itself. Energy plays a decisive role in all phenomena of classical and modern physics.

To start with, we have to concern ourselves with work. The result of interaction caused by a force is called work. This result becomes visible as motion, which can be described mathematically. But to bring about motion some resistance has to be overcome (Fig. 2.3). This can be the inertia of a mass, the attraction of a different body, friction or resistance against deformation. Generally work is calculated as

$$W = Fs \; [\text{Nm}] \text{ or } [\text{J}] \text{ or } [\text{Ws}] \text{ or } \left[\text{kgm}^2/\text{s}^2\right] \qquad (2.1)$$

Work is the product of force (F) and distance (s).

Fig. 2.3 Work

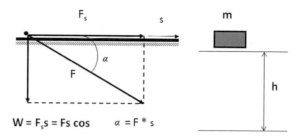

$$W = F_s s = Fs \cos \quad \alpha = F * s$$

If one wants to lift a mass gravity attraction has to be overcome:

$$W = - mgh \tag{2.2}$$

with h being the height to which the mass is lifted. In this case, we talk about elevating work. Another example is the work of acceleration.

The effect of work, however, can only be measured, when taking into account a certain amount of time—as work per unit time:

$$P = W/t \text{ [J/s] or [Watt] or } \left[kgm^2/s^2 \right] \tag{2.3}$$

P is called power. And this leads us to energy.

Once one has positioned a body at a certain height by the work necessary to achieve this, the position of this body carries in it the possibility to drop down again from this position, for example, from the edge of a diving board. During this process, while the body during its fall covers the same distance as the height, to which it had been lifted, the same amount of energy is released again, which had been invested in work to lift it in the first place. This energy could be brought to some practical use by, for example, using the impact of the body to crush other materials. As long as the body would remain at its height position it still contains the potential to release energy, if someone decides to unblock it. In this case, we talk about Potential Energy:

$$E_{pot} = mgh \tag{2.4}$$

The counterpart to this is called Kinetic Energy, also called momentum, which is released during actual motion:

$$E_{kin} = mv^2/2 \tag{2.5}$$

This formula can be derived in the following way:

$$E_{pot} = Fs \tag{2.6}$$

In the case of elevation energy F corresponds to mg and h to s. F again is am and $a = v/t$. If we assume that in our example the initial speed for the free fall $v_0 = 0$ and any average speed in between is designated by v, we obtain v at a point in time t:

$$v = (0 + at)/2 = at/2 \tag{2.7}$$

From this the resulting distance s equals:

$$s = v\,t = at^2/2 \tag{2.8}$$

By replacing ma by mv/t in (2.6) and a in (2.8) by v/t, we obtain for the kinetic energy:

$$E_{kin} = (mv/t)(v/t)t^2/2 = mv^2/2 \tag{2.9}$$

2.2.3 Temperature

A further quantity, which is of importance in thermodynamic considerations, is temperature. By our tactile sense, we are able to identify relative temperature differences and determine whether an object or a gas is warm or cold. In this case, our sensation does not only depend on the temperature itself but partially also on other determining factors like wind speed for example.

To arrive at an exact description we will start by looking at thermal equilibrium (Fig. 2.4).

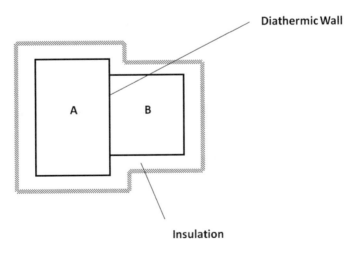

Fig. 2.4 Thermal equilibrium

Both containers are filled with gas at a certain pressure in a certain volume. The wall between both systems does not permit any exchange of substance or any electromagnetic effects. The gases in both containers are completely isolated from each other. We assume that in each container the temperature T is different. Furthermore, we assume that within each container there is thermal equilibrium. The steady temperature thus depends on volume and pressure:

$$T = f(V, p) \tag{2.10}$$

While each container by itself is initially thermally stable the total system is not in thermal equilibrium. The separating wall is transparent with respect to heat, and after some time the temperatures in both containers harmonize until the overall system obtains thermal equilibrium. The resulting temperature is different from both the original ones.

Experience shows that, if two systems are in thermal equilibrium with a third one, both systems are in thermal equilibrium with respect to each other. This is sometimes called the zeroth law of thermodynamics.

If we further consider two systems in thermal equilibrium and system A remains constant, but the pressure in system B p_B is varied, we obtain the results in Fig. 2.5.

These graphs are called isotherms. They represent in each case states of thermal equilibrium with system A depending on a third variable, the temperature (Eq. 2.10).

There is a precise relationship between pressure, volume and temperature, called the equation of state:

$$pV = mRT \tag{2.11}$$

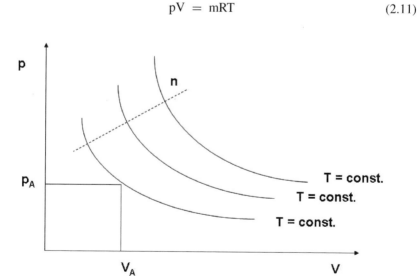

Fig. 2.5 Isotherms

This was first investigated by Robert Boyle. He derived the following law:

For ideal gases at constant temperature T and amount of substance n the volume
V is inversely proportional to the pressure p.

$$V_{T,n} \sim 1/p \text{ hence} \tag{2.12}$$

$$pV_{T,n} = \text{const.} \tag{2.13}$$

Or:

1. The volumes of one and the same amount of gas are inversely proportional
   to their pressures.
2. The densities of an amount of gas are proportional to the pressures but
   inversely proportional to their volumes.

Boyle's law is valid for ideal gases. But there are no ideal gases for any arbitrary
range of temperature and pressure.

Now, one could construct initially arbitrary systems and put them together with
other systems of a completely different kind to create thermal equilibriums in turn. If
the former display the temperature in any quantitative way they are called thermome-
ters. Depending on their design one has the possibility to develop a scale to be able to
compare temperatures. In everyday life in many countries, the Celsius scale is in use,
which is oriented along the phase states of water. If the temperature value is equal to
0 at the freezing point of water and the temperature value at boiling point is 100, one
has two fix points, which can be subdivided into 100 whole number fractions. These
are called degrees. In other countries, other scales have been developed historically:
Reaumur (France) and Fahrenheit (UK). All scales can be converted into each other,
whereby the Celsius scale has gained the widest distribution.

But now the freezing point of water (0°) is by no means the deepest temperature value, as we all know from winter times, when negative temperature values
are reported. One could think that the temperature scale would expand indefinitely
upwards and downwards. But this is certainly not the case for the downward region.
There exists a lowest temperature below which nothing goes anymore—in the true
sense of the expression. Its value is

$$-273.15\,^{\circ}C \tag{2.14}$$

Scientist do generally not use the Celsius scale but rather prefer Kelvin, the
absolute temperature, where

$$0 \text{ K corresponds to} -273.15 \text{ C} \tag{2.15}$$

The graduation of the Kelvin scale, however, corresponds to the Celsius scale. The value 0 K can never be reached in practice. Temperature values are dependent on the kinetic energy of atoms or molecules. The higher the kinetic energy the higher is also the temperature. However, at absolute zero there remain quantum fluctuations, which cannot be brought to a halt because of their nature. Thus absolute zero can be approached arbitrarily close, but in the end never reached completely.

2.2.4 First Law of Thermodynamics

Robert Mayer

Our considerations about potential and kinetic energy have shown that different forms of energy can be converted into each other. By this, the energy content does not change. However, systems exist, the energy state of which can be changed neither by kinetic nor by potential mechanical energy (or as will be shown later: electrical energy), but only by feeding heat into them for example. To be able to investigate the energy state of a system in general let us introduce the term "Internal Energy".

This term was coined for the first time by Robert Mayer, a physician, who, in the middle of the nineteenth century during his long journey to Batavia, took the opportunity to reflect on the relationship between the motion of the waves and the temperature of the water. Later on, he systematized his considerations leading eventually to the formulation of the first law of thermodynamics. However, he had to virtually fight until the end of his life for recognition, since he was just a medical but not a natural scientist.

Julius Robert von Mayer was born on 25 November 1814 in Heilbronn in Germany and died there in the same place on 20 March 1878. He became a surgeon and left Europe on 18 February 1849 on board the Dutch freighter Java for Batavia. During the transit, he considered for the first time the possibility of the warming of water caused by the motion of the waves. Furthermore, investigation of the blood of the sailors showed that at the end of a working day the oxygen content of the blood was less (it was darker) than in the morning. Oxygen had been burnt; the heat had been transformed into work.

During the journey back, Mayer had about 120 days for further reflections: his conclusions were: light, heat, gravity, movement, magnetism and electricity were all manifestations of the same "elementary force". Back home he drafted a first manuscript and sent it to Poggendorf's "Annalen der exakten Wissenschaft" (Annals of the Exact Sciences). It contained six pages of speculations. He also succeeded to calculate the mechanical heat equivalent. This is the amount of heat needed to increase the temperature of 1000 g of water from 0 to 1°.

His findings were refused by Prof. Noerrenberg in Tubingen and Prof. Jolly in Heidelberg. In 1842, Mayer sent a revised version of his treatise to Liebig's "Annalen der Chemie und Pharmazie" (Annals of Chemistry and Pharmacy). For seven years, he had no feedback from the scientific community. Further revisions were rejected by every publisher. Mayer then resorted to self-publishing and sent copies to every major scientific institution in Europe—without success.

In England, James Prescott Joule confirmed Mayer's propositions by experiment but omitted Mayer's name and reference in his publication. On the basis of Joule's findings, Hermann von Helmholtz published a paper called: "Ueber die Erhaltung der Kraft" (About the Preservation of Force): again no mention of Mayer. This led to a priority dispute.

Friends, acquaintances and his wife gave him up. He was admitted to psychiatric institutions and tried suicide. Later he resumed his medical practice. Many years later, the Royal Society announced that Robert Mayer had to be accorded the discovery of the law of the conservation of energy.

Deduction

Again we consider a closed system in a defined state of internal energy (Fig. 2.6). Such a system is called an adiabatic one, if its equilibrium can only be changed by executing work by it or on it. To advance an adiabatic system from a state of internal energy U_1 to a state U_2 the following work is necessary:

$$W_{12} = U_2 - U_1 \tag{2.16}$$

Let us now go back to our two containers A and B in Fig. 2.4. Together both constitute an adiabatic system, each container on its own, however, not. As we have already seen, the particular state of internal energy changes across the diathermic separating wall. This can also be formulated with reference to Fig. 2.7:

$$Q_{AB} = U_{A2} - U_{A1} \tag{2.17}$$

Fig. 2.6 Internal energy

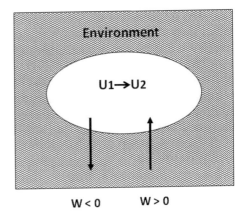

W < 0   W > 0

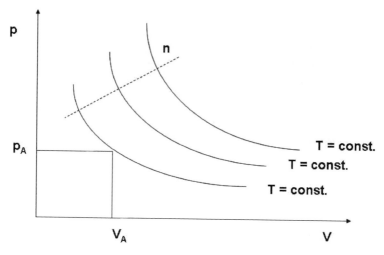

Fig. 2.7 Isotherms

$$Q_{BA} = U_{B2} - U_{B1} \tag{2.18}$$

However, for the combined systems

$$U_{A2} + U_{B2} = U_{A1} + U_{B1} \tag{2.19}$$

holds. This leads to

$$Q_{AB} = -Q_{BA} \tag{2.20}$$

We thus have to take into account an additional contribution for non-adiabatic systems when changing the internal energy.

$$W_{12} + Q_{12} = U_2 - U_1 \tag{2.21}$$

Q_{12} is called heat and is measured in [J] or [Ws] or [Nm]. This heat is energy, which appears at the boundary between two systems of different temperatures, and is exchanged between systems because of this temperature difference. Thus heat is a form of energy too. This leads us to the formulation of the first law of thermodynamics:

"In a closed system the total supply of energy, being the sum of mechanical, other and heat energy, remains constant."

The first law is called the principle of energy conservation. It is the basis for all further considerations in physics. Among other things, one of its consequences is the impossibility of a Perpetuum mobile. It also confirms: "there are no free lunches". That is, everything has its price, nothing is created out of itself but only by the transformation of something already existing into a different form.

2.2.5   Second Law of Thermodynamics

In thermodynamics, we differentiate between three types of processes:

- reversible,
- irreversible,
- impossible ones.

An impossible process would, for example, be the transfer of heat from a system of lower temperature to a system of higher temperature without external intervention. These kinds of processes will not be considered here any further.

A reversible process is defined the following way:

> "If a system, in which a certain process has been executed, can be brought back to its original state without leaving any changes in its surroundings, the process had been a reversible one."

Reversible processes are constructs, which are useful to calculate the efficiency of systems. These deliver maximum work, do not exist in nature or only in approximation, but serve as a reference for irreversible processes:

> "If the INITIAL STATE of a system, in which a certain process has been executed, cannot be recuperated without change in its surroundings, the process had been an irreversible one."

The second law of thermodynamics can be expressed qualitatively in this way:

> "All natural processes are irreversible."

During irreversible processes, energy gets sort of devalued. The result is loss of energy, which can be measured, however, against an idealized corresponding reversible process. One main reason for the irreversibility of processes is the occurrence of friction. But the second law gives further clues. It indicates the direction, in which thermodynamic and natural processes proceed. The directionality connected with this says that each point in time in the future is attached with higher entropy. To support these qualitative statements we need a state variable fulfilling the following conditions:

- It has to increase for irreversible processes.
- It has to decrease for impossible processes.
- It has to remain constant for reversible processes.

The state variable we are looking for had been introduced by Rudolf Clausius in 1865 and is called entropy. The definition for a change in entropy is

$$\Delta S = \int_{1}^{2} \frac{dQ}{T} [J/K] \tag{2.22}$$

The increase in entropy S is equal to the integral over the amount of heat input, lifting a system from state 1 to state 2, divided by the absolute temperature at which this happens.

Replacing dQ by the relevant energy equation, one gets

$$dQ = (dU + pdV) \tag{2.23}$$

We can thus state in summary:

1. Every system possesses a state variable S, entropy, the differential of which is defined by

$$dS = (dU + pdV)/T \tag{2.24}$$

with T the absolute temperature.
2. The entropy of an (adiabatic) system can never decrease. For all natural, irreversible processes entropy increases, for reversible processes it remains constant:

$$(S_2 - S_1)_{ad} \geq 0 \tag{2.25}$$

Entropy is also called a measure for the disorder of a system or the probability of a state. This means practically that an ordered system without external influence (adiabatic) always approaches a state of higher disorder. Together with this, the information about the original ordered state of the system gets lost. This is the course in nature (the decay of a dead organism) and in human history. To keep or create a state of higher, external energy has to be added. But this again only happens through other irreversible processes, which in turn generate energy loss. In the whole cosmos, entropy increases continuously.

2.2.6   Heat Conduction

Let us return to the graph with the isotherms.

This figure represents a temperature field. Along the isotherms, we find instances of the same temperature. The ratio between the respective temperature and the perpendicular distance Δn of the isotherms is called a temperature gradient.

$$\Delta T / \Delta n = \text{grad T} \tag{2.26}$$

In case of a temperature drop, heat spreads within a medium only in the direction of the temperature drop. The amount of heat transferred is called heat flow Φ in [J/s] or [W], the heat flow per unit area is called heat flow density in [W/m^2].

Experimental investigations by Jean Baptiste Fourier have shown that the amount of heat transferred is proportional to the temperature gradient, time and the cross-section perpendicular to the direction of the heat flow. The resulting relationship is

$$\Phi = -\lambda \text{grad T} \tag{2.27}$$

λ is called the coefficient of heat conductivity in [W/mK]. It is specific for different materials and depends also on further variables of the respective material: temperature itself, pressure, humidity, etc. In reference works, one would thus not find distinct numbers but rather graphs expressing these dependencies for the respective material.

Phases

In the sections above we have considered different phases of matter or materials, without invoking them explicitly each time: solid bodies, liquids and gases. For the latter two, we know equations of state depending on temperature. The systems we have looked at had always been subject to some simplifications: for example, systems with homogeneous phases, i.e. either solid or liquid or gaseous. But there exist of course many processes in nature not fulfilling these conditions. A simple process in that respect is to bring water to a boil. During this process, a state is reached at some point, where the gaseous phase emerges at the same time as the liquid phase. As such we have a system with heterogeneous phases. The description of such processes is sufficiently complex to go beyond the scope of this book.

Heat Exchanger/Cooling Systems

There are three main kinds of heat exchange:

- Direct heat exchange is a combination of heat and material transmission for separable mass flows, like a cooling tower.
- Indirect heat exchange, in which mass flows are spatially separated by a diathermic wall, like a motor car radiator.
- Semi-direct heat exchange via heat reservoirs. The heat reservoir is alternately heated by the hotter medium and thereafter cooled down by the colder medium to transfer thermal energy.

The efficiency of heat exchange depends on the geometry of the application. We differentiate between three basic designs:

- Counter flow,
- Co-current flow,
- Cross flow.

The complete heat exchange succumbs to Fourier's law as already mentioned above, but here somewhat more precise:

$$Q = \lambda F z \, dt / dx \qquad (2.28)$$

Here it says that an amount of heat Q flows through a cross section F of a solid body in z hours with the temperature gradient dt/dx of the heat flow; λ is again the coefficient of heat conductivity in [W/mK]. Heat transfer through a plane wall can be calculated as follows:

$$Qx = \frac{\lambda}{\delta} F(tw1 - tw2) \qquad (2.29)$$

with

$$\delta/(\lambda F) = R\lambda \qquad (2.30)$$

the thermal resistance.

2.2.7 Exergy and Anergy

According to the first law of thermodynamics there exists no process, by which energy can be created or annihilated. There are only transformations of energy from one form of energy to another. For these energy transformations, the balance equations of the first law are always applicable. But these do not contain information, whether a particular energy transformation is possible at all.

The second law of thermodynamics puts us in front of the following facts: energy can appear in different forms. Some of them can be transformed into other forms, for example, mechanical energy (kinetic as well as potential), including work. This is also true for electrical energy. We can summarize these types of energy under the term "exergy". If we assume a reversible process (which does not exist in nature) such energy transformations would be complete. At the same time, a limited transformation into internal energy or heat would be possible by other (irreversible) processes. Internal energy and heat, however, are types of energy, which are only transformable with restrictions. They cannot be transformed entirely into exergy. Taking the second law of thermodynamics into account these possibilities depend on

- the type of energy,
- the state of the energy carrier,
- the environment.

It follows that the total energy present in any environment cannot be transformed into exergy. The same is true for heat at ambient temperature. All this can be expressed the following way:

The complete energy content of all systems being in thermal equilibrium with the environment cannot be transformed into exergy.

Thus we can distinguish three TYPES of energy by using their degree of transformability as a criterion:

1. Unlimited transformable energy (exergy) like mechanical and electrical energy for example.
2. Limited transformable energy like heat and internal energy, the transformation into exergy of which is severely restricted by the second law.
3. Non-transformable energy like the internal energy of the environment for example, the transformation of which into exergy is impossible according to the second law.

Concluding the following definitions are valid:

Exergy is energy, which can be transformed in any other form of energy in a given environment; anergy is energy, which cannot be transformed into exergy.

$$energy = exergy + anergy \qquad (2.31)$$

By using these two "energy portions", we can now describe any balance for all energetic processes, since they form a unique relationship. Thus there are portions, which can be transformed in part into exergy, i.e. usable energy, and on the other hand portions, which are not available for transformation, i.e. utilization. Es already expressed explicitly in Eq. 2.31: energy types are composed of an exergetic and an anergetic portion. In some cases, one or the other portion can be equal to 0. Examples:

- electrical energy: anergy $= 0$,
- environment: exergy $= 0$.

It was the second law of thermodynamics which lead us to the classification of exergy and anergy. The existence of the concepts of exergy and anergy itself can be regarded as an alternative formulation of the second law. They are based on observations of nature. Against this backdrop, the second law can also be formulated the following way:

Energy is fundamentally composed of exergy and anergy with the possibility that one of the two can even be zero.

We have already been acquainted with the corresponding Eq. 2.31.

Now we can express the first law as well by using these two terms:

> The sum of exergy and anergy remains constant—independent from the process considered.

However, what applies to the sum of exergy and anergy cannot be applied to exergy or anergy alone. Let us consider once again the types of processes in this context. We can draw the following conclusions:

1. In all processes the sum of exergy and anergy remains constant.
2. In all irreversible process exergy is transformed into anergy.
3. Only in reversible processes exergy remains constant.
4. It is impossible to transform anergy into exergy.

These four statements determine the basis, on which energy transformations, such as we know from different heat and electricity generating facilities, function— independent of the type of energy transformation, be it wind power, nuclear power or photovoltaic cells. As we know already, all natural processes are irreversible. By this the available pool of transformable energy, i.e. exergy, continually diminishes, since portions of it get transformed into no further usable anergy. However, this does not concern the conservation of the sum of these two terms, expressed by the first law. Only the capability of utilization by transformation diminishes continually by the transformation of exergy into anergy itself. The logic of the four statements above is based on the definition of anergy (not convertible into exergy).

The conclusions of the second law about the behaviour of exergy and anergy can be proved the following way: The impossibility to transform anergy into exergy follows directly from the definition of anergy: it has been defined as energy, which cannot be transformed into exergy.

The ideal process to transform energy—a reversible process, which could be unwound backwards—is unfortunately only a thought experiment helpful for our considerations and useful as a yardstick but non-existent in nature and technology. This does not present a mere deficit, which could perhaps be eliminated by genial inventions or successive improvements. This limitation is not founded in our inability but is a fact of nature itself mathematically expressed by the second law of thermodynamics.

To put it simply: we are dealing both in nature and in our devices with some sort of "energy destruction" or energy devaluation. From a technical point of view, types of energy can be classified as follows:

> A form of energy is more valuable the higher its transformation potential, i.e. the higher its exergy component is.

Of course, this is considered having its later utilization in mind. Nature does not make such distinctions. Energy is energy, and its content does not change concerning its total sum. What we call "the loss of exergy", i.e. anergy, does not get lost according to the first law.

The whole discussion about energy does not turn around the total sum of energy itself at all, but rather specifically about the exergy portion. For every starting position, the point is to bring to use this portion: when heating, when generating electricity, in the application of power machines and in mobility. The energy required for these processes, however, is not its total sum but its exergetic portion, which can be applied usefully, i.e. transformed. Energy technology provides exergy. And at the end of the chain after utilization, what remains is anergy—at the consumer end after consumption.

Thus it would make more sense to talk about exergy sources. Nothing comes from the direct environment (anergy). In reality, the whole energy discussion should be called exergy discussion. Energy is not used up nor gets lost, and it can in no way be renewed (impossible process!). All considerations about optimization should lead in the direction to minimize exergy loss, i.e. anergy gain, during energy transformation. This starts already when selecting an energy source (high energy potential) and continues in process engineering and transformation technology. Thus notionally one could always try to approach a reversible process to preserve all original exergy under technical and economic aspects, although one has to be aware that in the end this cannot fully be achieved. In the end, the efficiency factor will tell, how close one has come to this ideal. Independently from purely technical considerations, economic aspects play an important role regarding cost minimization. This means that technical applications themselves are not the only criteria to solve certain challenges. Sometimes economic aspects could ask for more exergy loss that would be necessary under purely technical considerations.

2.2.7.1 The Calculation of Exergy and Anergy

To apply the terms exergy and anergy, we have to know their contributions to various energy forms. As already mentioned, electrical energy and some mechanical energy forms consist of pure exergy. From those energy forms, which are restricted in their transformation capabilities we will calculate the exergy and anergy portions with respect to heat and the mass flow as an energy carrier.

The exergy of heat is the portion of the heat, which can be transformed into any other form of energy, thus also into usable work. To calculate it, we imagine that the heat is put into a thermal engine, the working fluid of which is passing through a circular process. We obtain the exergy (E_{12}) of the heat (Q_{12}) as effective work (W_{eff}) and its anergy (A_{12}) as waste heat (Q_0) of this circular process. However, the effective work of the circular process only corresponds to the exergy of the heat input, if

- the circular process is executed reversibly (otherwise exergy is transformed into anergy, and the effective work is smaller than the exergy input), and
- the waste heat of the circular process is released only at the environmental temperature T_u, such that it consists only of anergy and corresponds exactly to the heat input.

The heat Q_{12} is transferred from some energy carrier to the working fluid of the thermal engine. During this process, the temperature of the energy carrier decreases from T_2 to $T_1 \leq T_2$, although it may also remain constant in case the energy carrier is a very large energy store. During the release of the heat Q_{12} (positive for the thermal engine) the entropy S_E of the energy carrier decreases as well:

$$Q_{12} = -\int_1^2 T dS_E \tag{2.32}$$

Figures 2.8, 2.9, 2.10 and 2.11 demonstrate these facts.

2.2.7.2 Heat Pump

One example of an exergy-anergy balance can be found in the so-called heat pump.

Heat pumps are employed to render waste heat from the environment utilizable for energy transformation. The classical scenario to transform thermal energy (exergy) into mechanical is the well-known propulsion of turbines by hot gases. The accompanying waste heat (anergy) is lost. The overall energy balance is

$$E_{ges} = E_{ex} + E_{an} \tag{2.33}$$

Heat pumps (Fig. 2.12) employ the inverse process. They pick up waste heat from the environment coming from different sources among them the ground earth. This

Fig. 2.8 Circular process

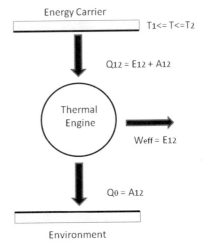

Fig. 2.9 Entropy decrease

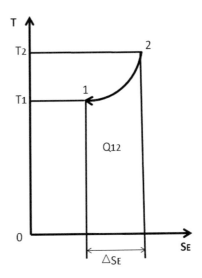

heat is used to evaporate a liquid with a low boiling point. Thereafter mechanical energy is added by compressing the vapor. Further on in the loop, the compressed gas is again expanded and condensed. The released heat can be used for heating purposes.

The corresponding energy balance is the following:

$$\dot{E} = \dot{E}an + |\dot{E}ex| \tag{2.34}$$

$$\dot{E}an = \frac{T_u}{T}\dot{E} \tag{2.35}$$

Fig. 2.10 Exergy

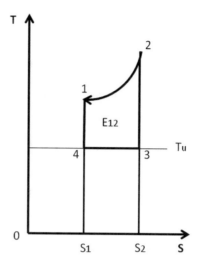

Fig. 2.11 Anergy

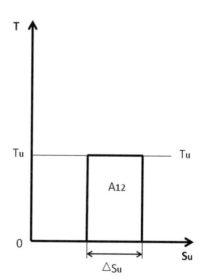

$$|\dot{E}ex| = \left(1 - \frac{T_u}{T}\right)\dot{E} \qquad (2.36)$$

with $\dot{E}$ the delivered heat flow, T_u the environmental temperature and T the delivered temperature.

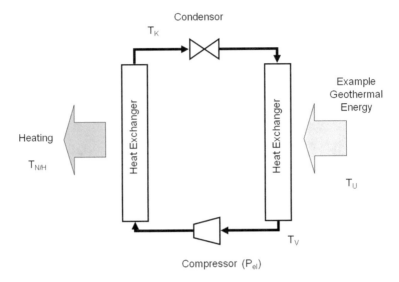

Fig. 2.12 Heat pump. Legend: T_u = Environmental temperature, T_V = Evaporator temperature, T_K = Condenser temperature, $T_{N/H}$ = Heating temperature

2.2.8 *Enthalpy*

The changes of state considered so far were related to stationary closed quantities of matter—closed systems. The work exerted by such systems was pure volume work. Contrary to these there are technologies like air compressors or piston steam engines using processes with a working medium passing through a container or a tube system. The flow through volume is called an open system, the process is a constant pressure process. It can be explained in an idealized fashion by the corresponding p, V-chart: a gas flows from a very big container with constant pressure p_1 via a valve O_1 (valve O_2 being closed) into a cylinder V and advances a piston K (Fig. 2.13).

The displacement work done corresponds to $-p_1 V_1$ and to the area under the isobar AB in the p, V-chart (Fig. 2.14). After closing the valves, the gas experiences a change of state, it expands to the volume V_2 and delivers the expansion work $-\int_{v1}^{v2} p \, dV$ corresponding to the area under the graph BC. The pressure sinks to p_2.

Fig. 2.13 Piston system

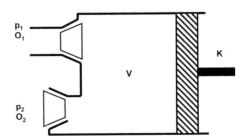

Fig. 2.14 p, V-chart

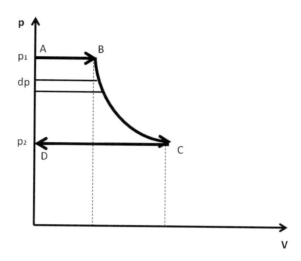

Thereafter the valve O_2 is opened and the gas is released into a big container with the same pressure. The required work is $p_2 V_2$. After closing O_2, the valve O_1 leading to the container with p_1 is again opened. The process starts anew with a new working medium. The machine works periodically. The total produced work can be used for technical purposes. Because of this, it is called technical work W_{tech} and corresponds to the overall area enclosed by the graph ABCD:

$$W_{tech} = -p_1 V_1 - \int_{v1}^{v2} p\, dV + p_2 V_2 = \int_{p1}^{p2} V\, dp \qquad (2.37)$$

Because of $dW_{tech} = V dp$, the first law can be shaped in the following way:

$$\begin{aligned} dQ &= dU + p\, dV - V\, dp \\ &= dU + d(p\, V) - V\, dp \\ &= d(U + p\, V) - V\, dp \\ &= dH - dW_{tech} \end{aligned} \qquad (2.38)$$

with the new state variable

$$H = U + pV \text{(Enthalpy)} \qquad (2.39)$$

The product pV is called displacement work. For isothermic changes of state, the technical work W_{tech} is equal to the volume work because of $p_1 V_1 - p_2 V_2$.

Enthalpy is a state variable, the difference between the initial and final state of which is equal to the heat exchanged at constant pressure.

2.2.9   Energy Balances in General

We have learned that the first and second law of thermodynamics are the main basis for the formulation of an energy balance. On this basis and by employing the concepts of exergy, anergy and enthalpy balances can be calculated for specific components or even whole facilities. From the point of view of physics, the results center on efficiency factors and the loss of usable energy.

2.2.9.1   Exhaustive Energy Balances

Be it energy or pollutants, for a fundamental evaluation of a specific energy form and its deployment, it is not sufficient to look at the final stage of utilization. An honest balance has to comprise all preproduction costs and all follow-up costs.

The following example was published in a German newspaper in 2019 [1]. It contains the summary of calculations by K. Radermacher and A. Herrmann:

The German Railways is known to be an environmentally acceptable means of transportation. This is certainly true for its operation in comparison to other means of transport. How does its infrastructure look like? The authors considered the ICE track between Cologne and Frankfort with a length of 170 km.

They used about 700 km of rail with production costs of 84,000 t of CO_2. Steel was imported from India. The use of 0.5 Million cm^3 of concrete generated 137,500 t CO_2. Further burdens were 30 tunnels, masts of the overhead wiring, 1300 suspenders, wiring, ropes and cables.

Conclusion: Before the first train rolled out, between 1995 and 2002 several million tons of CO_2 were generated.

Problem 1: Compilation of an Exhaustive Energy Balance

If one wants to establish an exhaustive balance comprising environmental soundness and economic viability the chain of calculation has to start early on. Compile a list of all important steps in the overall energy cycle relevant to this balance from raw materials production until waste disposal for an energy utilization plant.

Chapter 3
Physical Fundamentals of Energy Utilization

Abstract This chapter contains the most important physical basics and laws relevant to energy utilization. It comprises electricity, magnetism, electromagnetism, fluid mechanics and gas dynamics, as well as atomic and nuclear physics including atomic and nuclear models and the relevant reactions.

3.1 Electromagnetism

3.1.1 Introduction

Electromagnetism is all about the phenomenon of the electrical charge, which cannot be described by means of classical mechanics. Once movement comes into the game we encounter the phenomenon of the electrical current. In this case, as well considerations about energy balance play a role, for example, potential energy in connection with electrical voltage.

We will turn to magnetism, which in the past had been a field of interest of its own. Its relationship to electricity had been formalized by Maxwell's equations.

3.1.2 Charge

A strong hint although not proof that matter consists of atoms was given by the enhancements of the kinetic gas theory by Maxwell and Boltzmann during the second half of the nineteenth century. The explanation of gas pressure and its increase with temperature by collisions of gas atoms or molecules and their increase of velocity with temperature as well as the explanation of heat conduction and internal friction of gases through energy transfer by the colliding atoms gave explicit indications about the atomic nature of matter.

In the end, this supposition and Faraday's laws about electrolysis lead to evidence about the existence of a so-called electrical elementary quantum. The perception was that any monovalent charged atom always carries the same elementary charge e independent from its mass. Such monovalid or multi-valid charged atoms are called ions. A free negative elementary charge is called electron.

Electrons are only present in the shell of the atom, while nearly the complete atomic mass is bound in the atomic nucleus. This means that electrons can only possess a very small mass. Free electrons can be generated from metal surfaces or be irradiated using the photoelectric effect. Once there are free electrons there exists no further obstacle to determine their mass and exact charge.

The charge of a single electron has been determined to be

$$1.602176 * 10^{-19} \text{ [Coulomb]} \tag{3.1}$$

It has been convened that the charge of an electron should be negative, since in nature there exists also an opposite charge defined to be positive. It can be found with the proton, for example, one of the main constituents of the atomic nucleus.

The mass of the electron was found to be

$$m_e = 9.109382 * 10^{-31} \text{ [kg]} \tag{3.2}$$

Similar to gravitation with respect to the mass of a body in the case of electricity the charge of a body exerts a force. In this case, however, this force can act in two opposite directions—or not at all:

- attracting for bodies with opposite charge signs,
- repelling for bodies with the same charge sign,
- neutral (not counting mass attraction) for a body without charge and another one with any kind of charge.

The force exerted between two charged bodies can be calculated as follows (Coulomb's law):

$$\mathbf{F_c} = (1/(4\pi \varepsilon_0)) * \left((q_1 * q_2)/r^2\right)\mathbf{e_r} \tag{3.3}$$

or scalar:

$$F_c = (1/(4\pi \varepsilon_0)) * \left((q_1 * q_2)/r^2\right) \tag{3.4}$$

with the Coulomb constant

$$k_c = 1/4\pi \varepsilon_0 = 8.9875 * 10^9 \text{ [Vm/As]} \tag{3.5}$$

in vacuum, q_1 and q_2 being the respective charges, r the distance between the charge midpoints and $\mathbf{e_r}$ the unit vector along the connection between the charge midpoints.

Astonishingly in this case as well the force depends on the square of the distance. Similarly just as a mass creates a field of gravity around itself an electrical charge creates an electric field around itself. This will be discussed further down. Here are some additional principles concerning charge:

Two exchangeable charges are equal to each other, when a third one in the same arrangement exerts the same force on both of them one after another.

In a closed system the sum of positive and negative charges remains constant.

3.1.3 Current and Tension

Electrical current is the movement of electric charges or: the electrical current I tells us, how much charge Q is traversing an electrical conductor per unit time (Fig. 3.1):

$$I = Q/t \ [A] \tag{3.6}$$

To create electrical current at all one needs a current source. When using the same circuit and varying the source, a dependency is created between the measured intensity of the current and the intensity of the source. To describe the power of a source, a term called electrical tension U [V] is introduced. Electrical tension is a term that tells us how much energy is needed to induce a charged particle to move in an electric field.

From the example of a simple circuit, we deduce the proportionality between electrical current and tension:

$$U \sim I \tag{3.7}$$

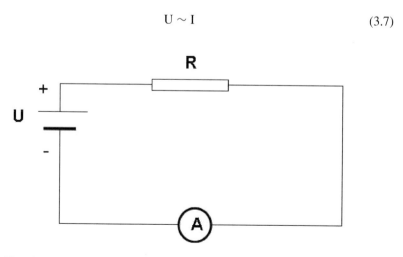

Fig. 3.1 Electric circuit

The proportionality factor is called electrical resistance R of a conductor. It is in turn dependent on the length l and the cross-section A of the conductor of a specific material:

$$R = \rho l / A \ [\Omega] \tag{3.8}$$

with ρ [Ω m] the specific resistance of the material, such that

$$U = R * I \tag{3.9}$$

Resistances are elements of electrical circuits. There are still others, of which we will only look at capacitors here. For this, we have to introduce the term of electrical capacity. First of all, there exists a proportionality between the charge of an electrical conductor and the tension:

$$Q \sim U \tag{3.10}$$

Equation 3.10 is valid for any configuration of isolated assembled conductors. The corresponding proportionality factor C is called the capacity of a conductor, such that

$$C = Q/U \text{ in } [C/V] \text{ or } [As/V] \text{ or } [F] \text{ (Farad)} \tag{3.11}$$

C depends on the geometry of the assembly and the dimension of the conductor.

Let us imagine the following assembly: an electrical circuit is interrupted by two conductor plates positioned opposite to each other with a non-conducting air gap d in between. Between these two plates, a tension U builds up and thus an electric field with charges of opposite sign on each of the opposing plates. The strength E of this electric field can be calculated as follows:

$$E = U/d \tag{3.12}$$

When A is the area of the conducting plates the corresponding charge is calculated as

$$Q = \varepsilon_0 EA = \varepsilon_0 AU/d \tag{3.13}$$

and thus the capacity of this empty plate capacitor by

$$C = \varepsilon_0 * A/d \tag{3.14}$$

Capacitors are used for the storage of electrical charges.

3.1.3.1 Direct Current

The intrinsic resistance of a conductor can be represented symbolically by a small rectangle. At the same time, this symbol is also used in circuit diagrams for other resistances of any kind like those visible in circuit logic and in practice on printed board assemblies. If there are several resistances in a single circuit, how do they enter into our tension Eq. (3.9)? The corresponding rules depend on the arrangement of the resistances and are based on Kirchhoff's Laws, which describe the branching out and the junction of current circuits:

- in series or
- in parallel (Fig. 3.2).

The law for series connection is

$$R_{ges} = R_1 + R_2 + R_3 \tag{3.15}$$

In a parallel circuit, the tensions in all resistances are equal.

The first Kirchhoff rule says that the sum of the currents, which converge on a branching point equals the sum of the out flowing currents:

$$I_{ges} = I_1 + I_2 + I_3 \tag{3.16}$$

The currents in our parallel connection are reciprocal to the resistances:

$$I = U/R \tag{3.17}$$

From this, it follows for the resistances:

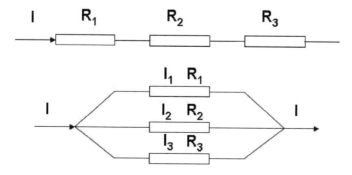

Fig. 3.2 Resistive circuits

$$1/R_{ges} = 1/R_1 + 1/R_2 + 1/R_3 \qquad (3.18)$$

Similar rules exist also for capacitors (in alternating currents) (Fig. 3.3). For series connections they are

$$1/C_{ges} = \sum_i 1/C_i \qquad (3.19)$$

and for parallel connections:

$$C_{ges} = \sum_i C_i \qquad (3.20)$$

Our considerations so far have implied only direct current, i.e. that kind of current, which is obtained uniformly from some voltage source. This is the occasion to comment on electricity generation as such.

Earlier on, the term electrolysis was introduced. Current flows—as we all know—not only via a metallic conductor such as copper wires but also through so-called electrolytes such as acids, bases or salt solutions. If one submerges a metallic conductor, for example, a copper plate, into an electrolyte, the metallic conductor develops a tendency to dissolve. But neutral atoms or molecules cannot dissolve, only ions. Ions are atoms lacking either one or more electrons or having one or more electrons in surplus to be neutral. So they can be charged negatively or positively. Their charge is a multiple of the electrons.

Fig. 3.3 Capacitive circuits

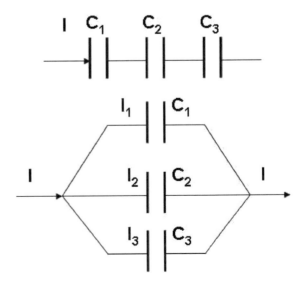

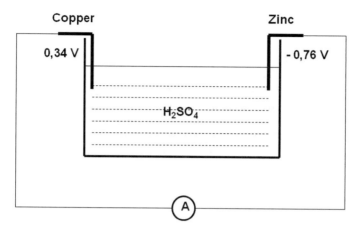

Fig. 3.4 Electrolysis

By the dissolution process, electrical tension is generated between electrolyte and conductor, which depends on the materials in question (Fig. 3.4). Now there exist—depending on the conductor material—either positive or negative tensions. For an example of copper or zinc in diluted sulphuric acid, they amount to:

Copper: +0.34 V
Zinc: −0.76 V.

If one submerges both materials at the same time one obtains a total tension difference of 1.1 V. From this tension difference one can now draw current. We have a primitive direct current battery.

Let us go back to our direct current circuits. When an electric current I passes through an Ohmic resistance R heat is generated, leading us to the subject of energy. This effect is used in various heating appliances. The amount of heat Q depends not only on the current, the voltage and the resistance but also on the time t. To generate heat electrical work W has to be exerted:

$$W = UIt \ \text{[Ws] oder} \ \text{[J]} \tag{3.21}$$

From this, we obtain for the power (work per unit time):

$$P = UI \ \text{[W]} \tag{3.22}$$

3.1.4 Magnetism

The term direct current indicates that there must be at least one more type of current. Before we can turn to alternating current a diversion to magnetism is required.

We all know magnets either from childhood toys or they meet us in everyday life, when we want to attach, for example, a poster to a pin board with a metallic core. Magnets possess two poles for attracting, for example, iron pieces (Fig. 3.5). If we hold two magnets in our hands, we realize that they attract or repulse each other depending on their orientation. Poles with the same orientation to the Earth's North Pole repulse each other, poles with opposite orientation attract each other. To attach a sign to a pole, we relate them to the North Pole. For a free magnet the pole, which orients itself in the direction of the North Pole of the magnetic field of the Earth is also called the North Pole and gets a positive (+) sign. The other one then becomes the South Pole ($-$). Contrary to the individual charges of electricity there exist no magnetic monopoles. Magnets appear only as dipoles.

3.1.4.1 Electromagnetism

Already very early on, people realized that electricity and magnetism showed similarities despite their individual peculiarities. For example, both forces were based on the existence of negative and positive polarities. Furthermore, it was observed that magnetisms and electricity could influence each other. Thus, for example, a conductor, when passed through by a current, creates a circular magnetic field around itself. If r is the distance between the conductor and some point P then the magnetic field strength H at P is

Fig. 3.5 Magnetism

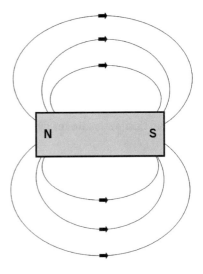

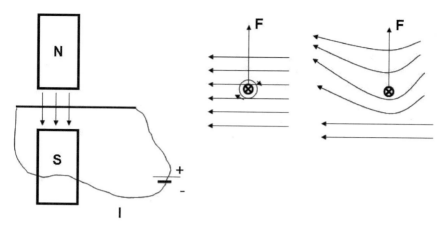

Fig. 3.6 Electric conductor in a magnetic field

$$H = I/2\pi r \; [A/m] \tag{3.23}$$

The magnetic field strength can be interpreted in analogy to the voltage of an electrical source. Inversely an existing magnetic field exerts a force on a conductor through which a current passes, when brought into the field (Fig. 3.6).

In this case, we are dealing with two superposed magnetic fields:

1. the field of the existing magnet,
2. the field created by the conductor current.

The concentric field intensifies the existing force lines as soon as they run into (almost) the same direction; it weakens the same as soon as the latter run into (almost) opposite direction. From this, we obtain a force perpendicular to the direction of the original magnetic field as well as perpendicular to the direction of the current, which tries to push the conductor into the direction of the weakening field. In this case, the so-called Left-Hand Rule is applicable:

"If one positions the left hand in such a way that the force lines enter the palm and the fingers point in the direction of the current, the thumb indicates the direction of the force."

This force can be calculated as follows:

$$F = BId \tag{3.24}$$

with d the length of the conductor, I the current and B the magnetic induction or magnetic flux density in $[Vs/m^2]$ or $[T]$ for Tesla. B, in turn, can be derived from:

$$B = \mu_0 * H \text{ [T]} \tag{3.25}$$

μ_0 is called induction factor and amounts to 12.566371×10^{-7} [Vs/Am]. B is a concrete measure for the number of force lines passing through a unit area. We will come back to magnetic induction, when we will discuss its electrical counterpart.

3.1.4.2 Induction

The motion of a conductor (length d) with the velocity $v = s/t$ through a magnetic field perpendicular to the force lines generates an electric tension, which in turn generates a current I in a closed circuit. With this electrical work is generated, which has to correspond to the added mechanical work:

$$Fvt = BIdvt \tag{3.26}$$

From this it follows:

$$U = Bvd \tag{3.27}$$

The corresponding Right-Hand Rule is

"The fingers point into the direction of the current, when the thumb points into the direction of the motion and the force lines enter the palm of the hand."

All electrical generators and motors rely on these induction phenomena.

3.1.5 Alternating Current

On the one hand, we have learned how to generate direct current, on the other hand, that current can also be generated by the motion of a conductor in a magnetic field. The latter phenomenon is used to generate alternating current. The simplest way is to turn a conductor coil in a homogeneous magnetic field—the principle of a dynamo or generator (Figs. 3.7 and 3.10).

The frequency of the revolutions is measured in Hz: 1 Hz corresponds to one revolution per second. Alternating current in most countries has a frequency of 50 [Hz].

However, alternating current does remain constant during one revolution, but changes with the angle of the coil with respect to the direction of the magnetic force lines. The generation of the current follows a sine curve (Fig. 3.8): according to this rhythm, the current initially increases and then decreases and so forth.

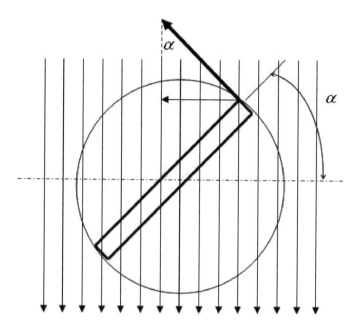

Fig. 3.7 Generator concept

Fig. 3.8 Alternating current
and alternating voltage

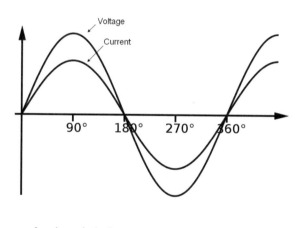

$$i = i_{max} \sin(\omega t) \qquad (3.28)$$

ω is also called angular frequency. The voltage taken at the coil clips follows the
motion in phase:

$$i = (u_{max}/R) * \sin(\omega t) \qquad (3.29)$$

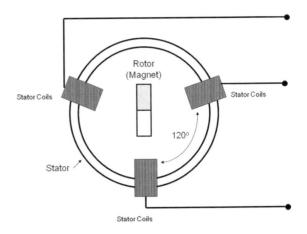

Fig. 3.9 Three-phase alternator

Everyday usage ignores the changes of alternating current and tension over time but refers to active values. In this case, the active value for alternating current corresponds to that value of direct current, which is required to generate the same power ($P = UI$ or $P = RI^2$). For the voltage it is analogous.

A special type of alternating current can be found in the three-phase alternator. With this device, three independent alternating voltages are induced. The necessary coils on a stator are offset by 120°. The stator is built from layers of electrically isolated magnetic iron sheets to minimize losses by eddy current (s. Sect. 3.1.6). The stator coiling is wired between the radially oriented poles of the stator in notches parallel to the axis. During the revolution of the rotor, the magnetic fields move along and cut the stator coils across the air gap (Fig. 3.9). By this, alternating voltage is generated with the three voltages having a phase shift of 120° each at the same frequency (synchronous with the revolution speed) and amplitude. The wiring is either done in star or delta manner (Fig. 3.10). Star wiring is used, when a neutral conductor is needed.

To be able to calculate with comparable values an active value is defined for the alternating voltage:

$$U_{eff} = U_s/\sqrt{2} \tag{3.30}$$

with U_s the peak voltage.

3.1.6 Maxwell's Equations

The Scottish physicist James Clark Maxwell was born on 13 June 1831 in Edinburgh and died on 5 November 1879 in Cambridge. He developed a set of four equations, later named after him (Maxwell's Equations). Furthermore, he discovered the speed distribution of gas molecules.

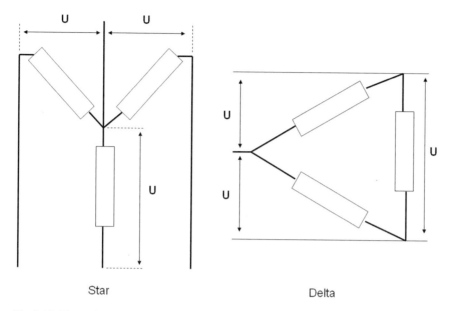

Fig. 3.10 Three-phase circuit

For our further considerations, we introduce a new term—the electrical flux density:

$$D = dQ/dA \ [As/m^2] \tag{3.31}$$

with dQ the alteration of charge over an area element dA. Let us now turn back again to our classical capacitor with no direct current passing across its air gap—other than with alternating current with a continuous periodic change of direction in such a way that the capacitor surfaces are charged and discharged permanently. The alteration of the flux density is called displacement current density:

$$j = dD/dt \tag{3.32}$$

Maxwell realized that—just as normal current—displacement current also generates a rotational magnetic field in its surroundings. His first equation for vacuum reads as follows:

$$\int Hdr = d/dt \int D * dA \tag{3.33}$$

Expressed in physics terms this means:

Every alterable electric field over time generates a rotational magnetic field.

Let us now imagine a ring-shaped conductor in a magnetic field, which alternates over time. Because of the law of induction, an electric rotational field with the field strength E_{ind} is created in this way:

$$\int E_{ind} \, dr = -d/dt \int B dA \tag{3.34}$$

– the second equation of Maxwell's Equations, meaning in terms of physics:

Every alterable magnetic field over time generates a rotational electric field.

Maxwell's Equations demonstrate in an impressive way the unification of two initially different forces of nature—the electric and the magnetic—to electromagnetism.

3.1.7   The Distribution of Electrical Energy

Electric power can be transmitted either as high current and small voltage or vice versa as small current with high voltage. High currents entail high construction costs, high voltages high insulation costs. Since the latter are generally much lower than the first long-distance transmission are carried out only with high voltage. Alternating current can be transformed to high transmission voltages. It has to be increased with distance and power to be transmitted. In general, it has to be transformed down to working voltage in one or two steps before usage.

3.1.7.1   Transformer

A transformer is used to transfer voltages from a higher (or lower) level to minimize transmission losses. It consists of two coils mounted on an iron core (Fig. 3.11).

The coils have different numbers of windings: N_1 (primary coil) and N_2 (secondary coil). As a consequence, they also possess different inductions L_1 and L_2. If one now feeds this setup with alternating current both coils will be coupled via the magnetic flux. This means:

$$L_1/L_2 = (N_1/N_2)^2 \tag{3.35}$$

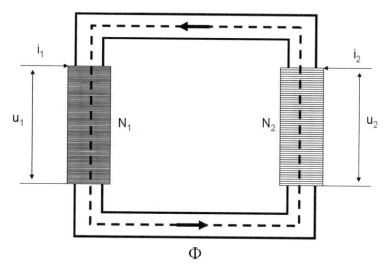

Fig. 3.11 Transformer

and for the alternating voltage

$$u_1 = U_0 e^{j\omega t} \tag{3.36}$$

and for the primary current (coil 1)

$$i_1 = u_1/(j\omega L_1) \tag{3.37}$$

This primary current now generates the magnetic flux

$$\Phi = L_1 i_1/N_1 = u_1/(j\omega N_1) \tag{3.38}$$

By the periodic alteration of the magnetic field, a self-induction voltage is generated against u_1 in the primary coil as well as the mutual induction voltage

$$u_2 = -N_2(d\Phi/dt) = -N_2(du_1/dt)/(j\omega N1) = -(N_2/N_1)u_1. \tag{3.39}$$

Since the phase of the magnetic flux Φ follows the phase of the primary voltage u_1 by $\pi/2$ and runs ahead of the secondary voltage u_2 by $\pi/2$ u_1 and u_2 are alternating voltages in paraphrase. They are transformed at the ratio of the number of windings:

$$R_T = \left| \frac{u_1}{u_2} \right| = N_1/N_2. \tag{3.40}$$

R_T being the transformation ratio.

3.2 Fluid Mechanics

3.2.1 Introduction

Every person who happens to have pumped up a bicycle tire must have sensed the resistance that the tire opposes against a continuous further pumping at the end of the operation. Everyone also knows that there is a relationship between this force and pressure—the pressure exerted by the compression of air—in any case by a medium, which is neither rigid nor point like.

In this chapter, we will consider forces and motions with respect to liquids and gases. But first of all, we have to define these media and their states of aggregation: when do we talk about a liquid and when about a gas? Concerning liquids, we will consider pressure (hydrostatics) but also buoyancy and other flow phenomena (hydrodynamics) and get acquainted with the most important equations of motion.

3.2.2 Liquids

3.2.2.1 Definition

Heuristically speaking the definition of a liquid presents no problem. Every child knows that there are three states of aggregation: solid, liquid and gaseous. However, there exist materials, where this classification is not unambiguous at all: tar, clay, glass and others. How do we define a liquid?

A liquid is a substance, which yields shear stress. To understand this, we have to introduce another term in fluid mechanics—tension (Fig. 3.12).

$$\tau = F/A \ \left[N/cm^2\right] \tag{3.41}$$

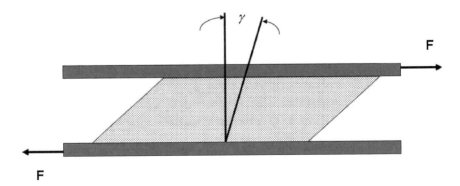

Fig. 3.12 Shearing

Tension is the ratio between force F and surface A. In our case, we talk about shear stress τ. A liquid deforms limitless in case shear stress is applied. Solid bodies to the contrary experience a limited deformation, which could decrease totally, partially or not at all, once the acting forces cease. The deformation of a liquid ceases once the acting shear strain subsides.

For our further definition of a liquid, we introduce the shear angle γ. It is obvious:

$$\gamma = f(\tau) \tag{3.42}$$

This means that the angle depends on the shear strain. This is certainly true for solid bodies. In case of liquids, there will be no shear angle at all, since shear increases limitless over time: the material flows, so that the measure of observation cannot be the shear angle but rather its speed of change:

$$\dot{\gamma} = f(\tau) \left[s^{-1}\right] \tag{3.43}$$

(3.43) is called the flow law of liquids.

However, there are different types of liquids. For our considerations, we will limit ourselves to the so-called Newtonian liquids. For those, the flow law is linear, so that we may replace (3.43) by:

$$\dot{\gamma} = \tau/\eta \tag{3.44}$$

The proportionality factor η in [Pa s] or [Ns/m^2] is called dynamic viscosity or dynamic shear viscosity, which is specific for each liquid. The behaviour of most liquids in everyday use (water, oil) can be described by this simplified model exactly enough. But even for these liquids, the constant does not remain constant but only specific. It can change with pressure and temperature. For everyday use, we may neglect the pressure dependency. Concerning temperature, there are appropriate tables in technical literature.

Most of what was mentioned in this context are valid for gases as well (with the exception of pressure dependency, which we have neglected here). At the moment, however, let us remain with liquids. Now we have to introduce the term of density:

$$\rho = m/V \left[g/m^3\right] \tag{3.45}$$

Density is the ratio between mass and volume in units [g/cm^3]. The density of droppable liquids is more or less independent of pressure and temperature.

3.2.2.2 Pressure

From what was outlined so far it follows that in a liquid at rest there is no shear stress. Forces acting upon a volume of the liquid via surrounding liquids or fixed walls are directed normally to the said volume of liquid. Furthermore, these forces are exclusive compressive and not traction forces.

The compressive force on an element of the surface of a liquid is proportional to the size of this surface element: In this way, pressure is defined as

$$p = F/A \ [Pa] \tag{3.46}$$

with 1 Pascal $= 1 \ [N/m^2]$.

Another consequence from our considerations is that pressure in a liquid at rest is exerted equally in all directions. The pressure originating from its interior weight is called weight pressure or hydrostatic pressure.

The hydrostatic pressure does only depend on the depth h and the specific weight ρ of the liquid. This means

$$P = h * \rho \tag{3.47}$$

3.2.2.3 Buoyancy

If one submerges a body of any shape in a liquid one observes a seeming loss of weight of this body. The reason for this is the Archimedean Principle.

It says that the amount, by which the weight of a body appears to have diminished, is equal to the weight of the amount of liquid replaced.

If V is the volume of the submerged body, ρ the density of the liquid and g the gravitational acceleration, the buoyancy force is

$$F_A = \rho * g * V \tag{3.48}$$

This can be illustrated by the fact that the compressive force, caused by the weight of the body, is equal to the force, caused by the weight of the displaced liquid on the total liquid. If the buoyancy of a body is greater than its weight when completely submerged, it will swim on the surface; if the buoyancy is smaller than the weight, it will sink. If the buoyancy is equal to the weight, it will float within the liquid.

3.2.2.4 Flow

To approach the motion of liquids let us introduce the term "liquid particle". Let us initially image a closed surface delimiting a volume of a liquid. This surface more or less floats along the flow with the liquid. If we diminish in thought this surface to an infinitesimally small point we can talk about a liquid particle. As such a liquid particle is an ideal entity. In the experiment, one can approach it again as a drop although in this case the infinitesimal size is of course already surpassed. Drops themselves have in reality their own internal dynamics.

We need this liquid particle to be able to introduce kinetic quantities. We know from mechanics that for example velocity or acceleration are quantities, which can only be deduced exactly for point-like entities.

The motion of a liquid is described by a velocity field:

$$\mathbf{v}(x, y, z, t) \tag{3.49}$$

with the components:

$$u(x, y, z, t), v(x, y, z, t), w(x, y, z, t) \tag{3.50}$$

This vector describes the velocity of a liquid particle in x-, y- and z-direction at a time t (Fig. 3.13).

Over time the directions described by $\mathbf{v}$ change. Those lines in a velocity field, the tangents of which correspond with the changing direction of $\mathbf{v}$ are called flow lines (Fig. 3.14). If the flow lines and thus the direction of $\mathbf{v}$ do not change over time at constant pressure and constant density the flow is called stationary.

One distinguishes flow lines and path lines (Fig. 3.14). Path lines describe the individual motion of a liquid particle over time. In stationary flows, flow lines and path lines are identical. Non-steady flows are, for example, whirls, in which liquid particles initially follow their individual path line with respect to an object moving in a liquid, before they swing into a flow line. For an observer moving with the disruptive object, the flow appears to be basically stationary.

Fig. 3.13 Flow lines

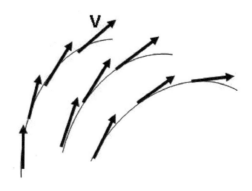

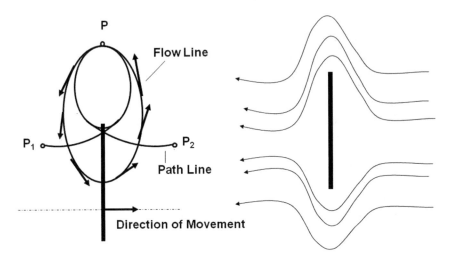

Fig. 3.14 Path lines

For a given problem, the consideration of a flat flow suffices quite often and (3.50) is reduced to

$$u = u(x, y, t), v = v(x, y, t), w = 0 \tag{3.51}$$

3.2.2.5   Equations of Motion

Apart from pressure, there is shear stress in a flowing liquid. Shear stress is provoked by friction at the confining walls among other things. For our following considerations, we will neglect this fact and concentrate on a frictionless stationary flow. For a liquid at rest, the following applies:

$$p_2 + \rho * g * z_2 = p_1 + \rho * g * z_1 \tag{3.52}$$

where p_1 and p_2 are pressures at two different points and z_1 and z_2 are the heights of these points. For a flowing liquid, we assume that between points 1 and 2 the pressure diminishes. But once the pressure diminishes, the velocity of the flow increases from 1 to 2.

The velocity component is u according to (3.51). It follows:

$$p_2 + u_2^2 * \rho/2 = p_1 + u_1^2 * \rho/2 \tag{3.53}$$

If we add gravitation to this liquid, we obtain Bernoulli's Equation:

$$p^2 + u_2^2 * \rho/2 + \rho * g * z_2 = p_1 + u_1^2 * \rho/2 + \rho * g * z_1 \qquad (3.54)$$

For a fixed point of reference and a single variable (3.54) changes to

$$P + u_2 * \rho/2 + \rho * g * z = C \qquad (3.55)$$

C may, however, vary from flow line to flow line. But for many flows, it is valid for the flow cross-section under consideration. Bernoulli's Equation expresses that the overall pressure (static and dynamic pressure) has the same value everywhere in a horizontal tube for example.

The equations for acceleration for stationary flows with their components read:

$$a_x = u * \partial u/\partial x + v * \partial u/\partial y + w * \partial u/\partial z \qquad (3.56)$$

$$a_y = u * \partial v/\partial x + v * \partial v/\partial y + w * \partial v/\partial z \qquad (3.57)$$

$$a_z = u * \partial w/\partial x + v * \partial w/\partial y + w * \partial w/\partial z \qquad (3.58)$$

dV being a small element of volume in a frictionless flow; a pressure force acts along the x direction:

$$-(\partial p/\partial x) * dV \qquad (3.59)$$

In addition, there is a volume force f_x * dv. In this case the mechanical basic law applies

$$\rho * dV * a_x = (-\partial p/\partial x + f_x) * dV \qquad (3.60)$$

After division by dV and insertion of (3.56) one obtains

$$\rho * (u * \partial u/\partial x + v * \partial u/\partial y + w * \partial u/\partial z) = \partial p/\partial x + f_x \qquad (3.61)$$

and analogous for the y- and z-directions:

$$\rho * (u * \partial v/\partial x + v * \partial v/\partial y + w * \partial v/\partial z) = \partial p/\partial y + f_y \qquad (3.62)$$

$$\rho * (u * \partial w/\partial x + v * \partial w/\partial y + w * \partial w/\partial z) = \partial p/\partial z + f_z \qquad (3.63)$$

(3.61–3.63) are called Euler's Equations of Hydrodynamics.

The Fountain of Sanssouci

"I wanted to have a fountain installed in my garden", Frederick the Great wrote to Voltaire in 1778, and continued: "…. Not a single drop of water arrived higher than fifty steps under the basin."

What happened? Frederick the Great wanted a fountain about 30 m gushing out of his well. The water should be taken from the nearby river of the Havel. However, the well was situated about 50 m above the Havel level. For the design, he had obtained the expertise of a scientist: Leonhard Euler. Euler's analysis was the starting point for the science of the hydraulics of non-steady flows in tubes. On the one hand, the results of his calculations were not applied at his time, on the other hand, his calculations are only applicable for frictionless flows, which do not exist in reality. Only later his equations were supplemented by a friction factor and became the Navier–Stokes Equation.

$$\rho(u\partial u/\partial x + v\partial u/\partial y + w\partial u/\partial z + \partial u/\partial t) = \partial p/\partial x + f_x + \eta \Delta u$$

with η the well-known dynamic viscosity.

Problem 2: Calculation of a Specific Weight

The weight of o body in the air is 27 g. The weight of the same body in water is 16.5 g. The loss of weight between air and water is $17 - 16.5 = 10.5$ g. The volume of the body is 10.5 cm³. Calculate the specific weight of the body.

Problem 3: Register Tons

Is there any relationship between register tons and the Archimedean Principle?

3.3 Gas Dynamics

3.3.1 Ideal Gas

The term gas dynamics delimits the following consideration to flow processes, in which the flow medium changes its state (density, temperature) significantly due to the flow itself. This will happen, when the flow comes close to the speed of sound. Although under very high pressure even liquids can adhere to this definition we will consider only gases in the following. Ideally, friction can be neglected, no energy is transferred by heat conduction and the flow is stationary.

3.3.1.1 Speed of Sound

The deduction of the equation for the speed of sound in gases can be done via an infinitesimal change of mass in a compressible gas. In this case one once again commences with Euler's Equations (3.21–3.23) by introducing a time-dependent term:

$$\rho \partial v / \partial t \tag{3.64}$$

The general equation of state for gases from thermodynamics reads

$$pV = mRT \tag{3.65}$$

with p the pressure, V the volume, m for mass and T temperature. R is the gas constant, which is specific for each gas with the unit

$$[J/(kgK)] \tag{3.66}$$

Furthermore, we require the so-called adiabatic exponent

$$\kappa = c_p / c_v \tag{3.67}$$

with c_p and c_v in [J/kg K] being the specific heat capacities of a gas at constant pressure or constant volume, respectively. The specific heat capacity corresponds to that amount of heat, which is necessary to raise the temperature of 1 kg of a substance by 1 K. We spare the exact deduction for the speed of sound by employing differential equations here. The result is

$$a = \sqrt{\kappa RT} \tag{3.68}$$

As a convention in gas dynamics, the flow velocity is expressed without dimension, by dividing it by the speed of sound. The resulting ratio is called Mach's Number.

3.3.1.2 Sound Propagation

We look at the three following cases (Fig. 3.15):

- gas at rest,
- flowing gas with u < a,
- flowing gas with u > a.

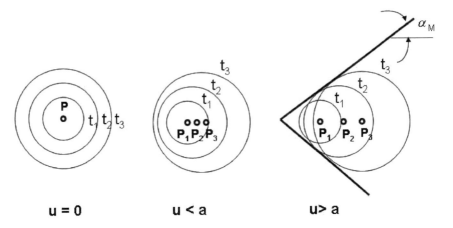

Fig. 3.15 Sound propagation

In a gas at rest, a small point-like pressure disturbance spreads uniformly with the speed of sound. P is the centre of the pressure disturbance. The spreading is done in concentric spheres similar to the effect, when one throws a stone in a pool of water. The gas particles remain at their fixed position, only the energy propagates.

In a gas flowing with a velocity below the speed of sound, the disturbance source travels with the flow. Concentric spheres are generated as well, the centers of which, however, shift downstream.

In a gas flowing with a velocity above the speed of sound (right in Fig. 3.15), the disturbance remains within a defined space cone, in which the disturbance source moves along. The angle of the space cone, within which the disturbance propagates, is called Mach's Angle. The relationship between angle and Mach's Number is

$$\sin \alpha_M = 1/M \qquad (3.69)$$

The acuteness of the angle increases with the velocity of a gas in the supersonic region.

3.3.2 Turbines

A characteristic of all steam turbines (Fig. 3.16) is the blading of the shaft and the housing. Steam is directed through the blades and thus forces the shaft to turn. The blades in the housing are called guide blades, those on the shaft rotor blades. Steam turbines are further distinguished according to the flow direction: in axial turbines, the steam passes the machine in parallel to its axis, in radial turbines in rectangular direction to the axis. Today's large turbines are exclusively axial turbines, since the

Fig. 3.16 Steam turbine (Tenerg, Bruenn)

size and thus the power of radial turbines is limited by the centrifugal forces at the tip of the blades.

Furthermore, steam turbines can be differentiated with respect to the eduction of the exhaust steam: condensing turbines direct their exhaust steam into a condenser, back-pressure turbines render the exhaust steam to a pipeline network for district heating. A third distinguishing feature is to be found in the pressure reduction of the steam: in constant-pressure turbines, the steam is relieved mainly in the guide blades, in reaction turbines, this happens equally in the guide and the shaft rotor blades.

Depending on the size and power of an axial turbine, the number of junctions at the turbine for live and exhaust steam increases. In this context, one differentiates between single-flow and multi-flow turbines. For large turbines, the overall turbine housing is divided into pressure stages. A typical version of steam turbines in a coal power station is made up of a single-flow high and medium pressure part, whereas the low-pressure part is realized as double- or even four-flow. Multi-flow not only increases a turbine's power, but also decreases the reaction of the shaft bearing in the direction of the shaft.

The ideal lossless steam power process runs as follows (Fig. 3.17):

- 1–2: frictionless and adiabatic pressure increase of the working fluid water up to the pressure prevailing in the steam generator.
- 2–3: heating of the water to the evaporation temperature for that pressure.
- 3–4: conversion of the liquid water to steam at constant pressure.

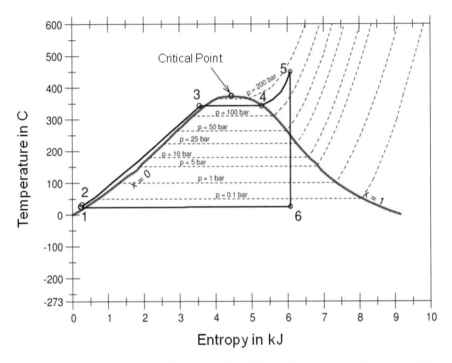

Fig. 3.17 Thermodynamic cycle of a steam turbine (Markus Schweiss; https://commons.wikipe dia.org/wiki/File:TS_Dampfturbine.png)

- 4–5: continued heating and superheating of the steam at constant pressure.
- 5–6: frictionless and adiabatic decompression of the steam at constant entropy, while at the same time first water droplets are created.
- 6–1: isobaric condensation of the wet steam in the condenser.

The area enclosed by the cycle process represents the technically usable work in relation to the amount of steam passing through.

3.3.2.1 Euler's Turbine Equation

If for a radial turbine, we designate the flow velocity relative to an observer by v, the absolute value of this velocity before entry into the rotor is v_1 and after exit v_2. Furthermore, v_1 has a component v_{u1} in the direction of the circumference of the rotor. When leaving the rotor this component shall be v_{u2}.

For our further considerations, we use the definition of angular momentum. If the distance of a mass point m from the shaft is r, then this mass point moves with a velocity w around the shaft. The plane, in which w moves, stands perpendicular to r. The angular momentum of the mass point is calculated as follows (scalar):

$$L = mwr \tag{3.70}$$

The corresponding torque is

$$M = dL/dt \tag{3.71}$$

If we apply, the calculation of the angular momentum of the rotor with the radii r_1 and r_2, connoting arbitrary distances from the shaft, at the time t, we obtain

$$dL/dt = \dot{m}(r_2 c_{u2} - r_1 c_{u1}) \tag{3.72}$$

with c_{u1} or c_{u2} being the respective velocity components of the inflowing or exhausting gas in the direction of the circumference of the blade wheel. The usable momentum of the turbine turns out to be

$$M = \dot{m}(r_a c_{us} - r_e c_{ue}) \tag{3.73}$$

with $\dot{m}$ the mass flow of the steam or the gas and r_a and r_b the exit or entrance radius of the turbine, respectively, where

$$r_a - r_e = b \tag{3.74}$$

is the blade width.

3.4 Atomic Physics

3.4.1 Introduction

After insights about the nature of electromagnetic radiation delivered the first important incentives for the development of a new theory—quantum theory—and soon after the pioneering work of Max Planck two other fundamental discoveries were made. They were widening the horizon beyond radiation to material particles themselves.

The so-called photoelectric effect had been known already around 1900 from different observations: among other things, it was known that it led to the liberation of electrons from metallic surfaces by light rays. However, a conclusive explanation was lacking until the year 1905. In classical terms, this effect had been explained by the assumption that the exposure rate of the incoming light (i.e. the amount of incoming light per unit area and time measured in [W/m^2]) would eventually force the electrons to leave the metal. However, this explanation did not conform to experimental findings.

Einstein's explanation was based on an assumption, which would turn the model of light as a pure wave, at that time universally accepted, upside down: his explanation was based on the fundamental supposition that light must indeed have particle qualities. These particles are today called photons. Thus the dilemma of the wave-particle duality had been revived—but this time not as a discussion of two different points of view but as real phenomena in nature itself.

This duality was in turn confirmed nearly twenty years later by scattering experiments by Compton. His results again could only be explained by the particle characteristics of electromagnetic waves.

That a radically new approach was inevitable was made obvious by the fact that matter itself, which had been assumed to consist of particles, showed wave characteristics in optical arrangements. The term "matter waves" was born. These findings made a revision of the classical view of the world inevitable. While at the beginning of the twentieth century, people were still convinced that electromagnetic waves—especially light—could be described adequately by their wave characteristics and matter particles like molecules, atoms or electrons be understood as particles idealized by a mass point, photo effect and Compton effect as well as matter waves saw to it that this view of the world was shaken.

3.4.2 Radiation

3.4.2.1 Thermal Radiation

From thermodynamics, we know that each body is in constant heat exchange with its environment—a process, which only in rare cases ceases, when it reaches some state of equilibrium. This process is an exchange of energy. It may happen by direct contact with another body, a liquid or a gas. In vacuum, heat exchange does not take place, but energy exchange happens by emission or absorption of electromagnetic radiation. In this case, we talk about thermal radiation. It is initially characterized by a continuous spectrum comprising all possible wavelengths (Table 3.1).

The wave length of visible light depends on the temperature of a body: the higher the temperature the shorter the wave length—from infrared to ultraviolet. According to the first law of thermodynamics, a hot body must cool down leading to emission, whereas a cold body absorbing heat will warm up.

3.4.2.2 Terms and Quantities of Radiation Physics

There are a number of terms relevant to our further considerations. Let us begin with radiant power:

$$\Phi_e = dQ_e/dt \quad [J/s] \text{ or } [W] \tag{3.75}$$

Table 3.1 Electromagnetic spectrum

Range	Type	Wavelength	Frequency
Audio frequency	Technical alternating currents	10^5 to 10^7	10 to 10^4
HF	LW	10^3 to 10^4	10^5
	MW	10^2 to 10^3	10^6
	SW	10 to 10^2	10^7
	VHF	1 to 10	10^8
Microwaves		10^{-3} to 1	10^9 to 10^{12}
Infrared radiation		10^{-6} to 10^{-3}	10^{12} to 10^{14}
Visible light		10^{-7} to 10^{-6}	10^{15}
Ultraviolet radiation		10^{-8} to 10^{-7}	10^{15} to 10^{16}
X-rays		10^{-12} to 10^{-8}	10^{17} to 10^{20}
γ-radiation		10^{-15} to 10^{-11}	10^{19} to 10^{24}

with Q_e the radiant energy. Thus radiant power is the change of radiant energy over time.

Normally a body radiates its energy spherically into space. It is therefore useful to relate radiant power to the solid angle constituted by a cone. From this we obtain the radiant intensity:

$$I_e = d\Phi_e/d\Omega \quad [W/sr] \tag{3.76}$$

with Ω, the solid angle to be calculated by the following way:

$$\Omega = A/R^2 \tag{3.77}$$

where A is the section of the surface of the sphere constituting the cone and R is the radius of the sphere around the radiating body. If one relates the radiant power to the area of the radiating emitter as seen by an observer, one obtains the radiation density:

$$L_e = d^2\Phi_e/(d\Omega x dA_1 \cos\alpha_1) \quad \left[W/(m^2 sr)\right] \tag{3.78}$$

With dA_1 an element of the surface of the radiation emitter and α_1 the angle between the surface normal and the receive direction. The radiation density depends on a given temperature on the wavelength λ. From this, we obtain the so-called spectral radiation density:

$$L_e\lambda = dL_e/d\lambda \quad \left[W/(m^3 sr)\right] \tag{3.79}$$

Besides its relation to the solid angle, the radiation flux density can also be a measure of the radiation flux relative to the emitting surface element:

$$M_e = d\Phi_e/dA_1 \; [W/m^2]$$ (3.80)

However, against this the irradiance on a receiver is defined as

$$E_e = (d\Phi_e/dA_2) \cos \alpha_2$$ (3.81)

With A_2 the element of the receiver surface and α_2 the angle with respect to the surface normal in the direction of the radiation flux. For us it is of interest that the radiation flux emanating from the sun and received on earth per m^2 at noon amounts to (solar constant):

$$E_{eS} = 1395 \; [W/m^2]$$ (3.82)

The energy density contained in a radiation field is calculated as

$$w = dQ_e/dV \; [J/m^3] \; \text{or} \; [N/m^2] \; \text{or} \; [Pa]$$ (3.83)

By conversion, the radiation pressure of an electromagnetic wave can be derived as

$$p = E_e/c$$ (3.84)

with c the speed of light.

3.4.3 Kirchhoff's Radiation Law

We have learned that there is emission and absorption. Besides these there exist two more modes:

- reflection,
- transmission,

i.e. a body can throw back (reflect) light or let it pass. All four exchanges depend on the temperature and surface constitution of a body, and can be expressed by a suitable measure.

The absorbance is defined as

$$a = \Phi_a/\Phi_0$$ (3.85)

with Φ_0 the total radiation flux and Φ_a the absorbed radiation flux. Before concentrating on further definitions, let us consider absorption in particular, since it will play an important role in the following section. There are materials like smut, which absorbs radiation nearly completely, in this case by 99%. A body, which absorbs all of the radiation impinging on it, and this for all wavelengths and temperatures, is called a blackbody. But now let us consider the other kinds.

The reflectance of a body is calculated as follows:

$$\rho = \Phi_r/\Phi_0 \tag{3.86}$$

with Φ_r the reflected radiation flux. If $\rho = 1$, the body is called a white body. The transmittance is defined as follows:

$$\tau = \Phi_{tr}/\Phi_0 \tag{3.87}$$

with Φ_{tr} the transmission flux. Contrary to the definitions given so far, which were based on the respective radiation flux, the emissivity is derived differently:

$$\varepsilon = L_e/L_{e(s)} \tag{3.88}$$

In this case, the radiation density is taken into account. Thus the emissivity is the ratio between the radiation density of a non-blackbody and the radiation density of a blackbody $L_{e(s)}$. Emission and absorption depend on temperature and wavelength. Experimental investigations have shown that for two different bodies the following relationships exist:

$$\varepsilon_1(\lambda_2 T)/a_1(\lambda, T) = \varepsilon_2(\lambda, T)/a_2(\lambda, T) \tag{3.89}$$

If one of them is a blackbody, the equation is

$$\varepsilon_s(\lambda, T) = a_s(\lambda, T) = 1 \tag{3.90}$$

From this follows Kirchhoff's Radiation Law:

$$\varepsilon(\lambda, T) = a(\lambda, T) \tag{3.91}$$

Independent from its other characteristics the emissivity of a body is equal to its absorbance at a given temperature and wavelength. This means that a body capable to absorb a lot of radiation can also emit the same amount of radiation.

3.4.4 Planck's Radiation Law

In 1900, Max Planck investigated a so-called blackbody. A blackbody is, as already mentioned, a body, which completely absorbs all radiation of any wavelength impinging on it. Furthermore, it continually emits radiation with a spectral energy distribution, which is independent on its own nature and else on its absolute temperature. The essential finding of Planck was that the spectral energy distribution of the emission of a blackbody is not continuous but happens in discrete frequencies. The emitted waves observed corresponded to quantized energy states according to the equation

$$E_s = Z * h * \nu \tag{3.92}$$

with v the frequency, Z a whole number and h a constant, which constituted the basis of all quantum physics—Planck's quantum of action:

$$h = 6.62507 * 10^{-34} \text{ [Js]} \tag{3.93}$$

This first transition from classical physics to atomic physics was followed by a further rupture. Until now we have assumed that radiation was a continuous phenomenon—independent from its rank in the spectrum. However, in the world of atoms, certain interactions between radiation and matter can only be explained, when radiation itself possesses some atomic characteristics. These considerations resulted in the explanation that radiation is absorbed or emitted by light quanta or photons. The photon energy corresponds to

$$E = h * \nu \tag{3.94}$$

Photons do not have a rest mass. They only exist once when they move with the velocity of light. The Theory of Relativity, however, leads to a relationship between energy and mass, and as a consequence, the photon also must have an inert mass depending on its frequency.

3.4.5 Light Measurement

When applying electromagnetic waves in the visible region, called light, the above definitions related to radiation are of little help. For illumination applications, therefore special photometric terms have been introduced. These we shall look at in detail. Let us commence with the luminous intensity analogous to the radiation intensity

$$I_v = d\Phi_v / d\Omega \quad \text{[cd]} \tag{3.95}$$

with Φ_v the luminous flux. The Candela unit [cd] has been defined as follows:

1 [cd] is the luminosity of a light source emitted in one direction with monochromatic radiation of 540×10^{12} Hz frequency and intensity of 1/683 [W/sr]. The luminous flux Φ_v is measured in [lm] Lumen:

$$1 \text{ [lm]} = 1 \text{ [cd sr]} \tag{3.96}$$

or 1 [lm] is the luminous flux emitted by a light source with a radiation intensity of 1 [cd] independently of any angle into a solid angel of 1 [sr]. In analogy to radiation density, there exists the luminance:

$$L_v = dI_v/dA_1 \cos \alpha_1 \tag{3.97}$$

and to irradiance the illuminance:

$$E_v = (d\Phi_e/dA_2) \cos \alpha_2 [lx] \tag{3.98}$$

in Lux, where a luminous flux of 1 [lm] generates an illuminance of 1 [lx] on an area of 1 m². Furthermore,

$$E_v = I_x \cos \alpha_2/r^2 \tag{3.99}$$

The latter is called the photometric law. It says that the illuminance diminishes with the square of the distance. The following relationship exists between two light sources emitting, although from different distances, but onto one and the same surface under the same incidence angel:

$$I_1/I_2 = r_1^2/r_2^2 \tag{3.100}$$

This means that the luminous intensity of two light sources with equal illuminance relates to each other as the square of their distances from the illuminated surface.

Table 3.2 shows the luminance of different sources.

And Table 3.3 gives the illuminances.

Table 3.2 Luminances (*Source* H. Stroppe, "Physik fuer Studenten der Natur- und Ingenieur-Wissenschaften", 2008, Munich)

Source	Luminance [cg/m²]
Fluorescence	$10^{-4} \dots 0.10^2$
Full moon	5×10^3
Candle flame	7×10^3
Tungsten coil in a light bulb	$10^7 \dots 3.5 \times 10^7$
Crater in a carbon arc lamp	1.6×10^8
Mercury high pressure lamp	$2.5 \times 10^8 \dots 10^9$
Sun	$10^7 \dots 10^9$
Xenon high pressure lamp	$10^9 \dots 10^{10}$

Table 3.3 Illuminances
(*Source* H. Stroppe, "Physik
fuer Studenten der Natur- und
Ingenieur-Wissenschaften",
2008, Munich)

Illuminance	Object
Moonless night	About 0.0003
Night with full moon	About 0.2
Sunlight in summer	10^5
Parking lot	3
Sidewalks, streets	5 … 20
Living quarters	40 … 150
Store rooms	50 … 200
Offices, class rooms	300 … 500
Measuring station	750
Micro-assembly	1000

3.4.6 The Photo Effect

The photo effect was first mentioned by Heinrich Hertz in 1886. It appeared as a side effect in his experiments with radio waves. His detector reported an increase of signals, when he irradiated his metallic radio sources with ultraviolet light. But Hertz did not pursue this phenomenon any further.

In 1888, Wilhelm Hallwachs was the next person to occupy himself with this phenomenon. For this purpose, he selected electrically charged zinc plates. He applied UV light and observed a loss of the negative charge of the plates.

A first breakthrough came about, when in 1899 Joseph J. Thomson provided evidence that under certain circumstances irradiated metal emitted electrical charges. The carriers of these charges were electrons, which he himself had discovered only in 1897.

Only Phillip von Lenard engaged in more detailed investigations in 1902. He varied the irradiation, the surfaced and the metals. For the first time discrepancies appeared between the generally accepted theory at that time and observations.

It was certain that irradiation by light of metallic surfaces liberated electrons. Other charge carriers with for example positive charge were not identified. Furthermore, it was proved that the number of electrons generated with this method increased with the radiation power: the more light hit a surface the more electrons were set free. Thus the photoelectric effect had been discovered.

Although the electron flux increased with radiation power no dependency between the latter and the energy or velocity of the electrons generated predicted by classical theory could be established: more light did not create more energetic or faster electrons. However, there exists some relationship between the color of the light, i.e. the frequency of the radiation used, and the liberated electrons. In the following, we will discuss the contradictions between past theory and experimental observations.

Theoretical Considerations

For a long time, it was assumed that light was a transversal electromagnetic wave. Its velocity would be c. Furthermore, it was assumed:

- The energy transported would be transferred to a certain number of electrons.
- The generation of free electrons was explained in this way: the metal contains free electrons induced into vibration by the electric field of the wave. These vibrations would build up until they could overcome the binding forces of their environment. By increasing the irradiance, the number of emitted electrons would grow as well as their kinetic energy (velocity).
- By diminishing the irradiance, however, a measurable delay of electron generation should be remarked, since the development of the necessary vibrations in the metal would require a certain amount of time.

Experimental Results

Experiments, however, delivered results contradicting the above theory:

- First of all, people had detected a lowest frequency ν_0. Below this cut-off frequency, no more electrons would be liberated—independent of the irradiance of the light.
- The proportionality between the irradiance and the number of electrons generated was confirmed but not the prediction about the dependency of the electron velocity on it. It was discovered that the velocity of the electrons depended on the frequency of the light given by ν minus the cut-off frequency ν_0. Irradiance did not play any role in this.
- Also, a time delay for the generation by radiation of lower intensity could not be observed. Apparently, the photo effect happened to be a spontaneous event.
- It could not be explained by the tools of classical physics.

The explanation of this phenomenon was delivered by Albert Einstein. He constructed the following relationship:

$$E_{kin} \text{ (electrons)} = (m/2) * v^2 = h * \nu - \Phi \tag{3.101}$$

with m the electron mass, v the electron velocity, ν the frequency of light, h Planck's quantum of action and Φ the so-called work of emission. The latter is a constant specific for every metal. The quantity 'h * ν' was in this case interpreted as an energy packet; a packet of light energy transferred to electrons. If the energy surpassed the work necessary to liberate the electrons, they could leave the metal while possessing the remainder of the energy packet as kinetic energy. This was the proof that light as well could appear as single quanta: photons. Thus it was demonstrated that electromagnetic waves can have particle characteristics.

In his interpretation, Einstein resorted to the conception of quantization of energy based on considerations by Max Planck concerning blackbody radiation. As we have seen above, Planck explained its energy distribution by assuming that electromagnetic radiation consisted of discreet quanta according to the relationship:

$$E = h\nu \tag{3.102}$$

with ν the radiation frequency and h Planck's well-known quantum of action.

Using this approach of Planck's quantum hypothesis Einstein concluded that a continuous distribution of light in space was not possible, but that light was always moving in discreet packets. With this, he went a step beyond Planck, who had introduced quantized packets only as theoretical constructs. The distribution of quantized light in space led to the conclusion that light must have particle characteristics. These particles were called photons by Einstein. Their energy can be expressed in the following way:

$$E_{Photon} = h\nu = hc/\lambda \tag{3.103}$$

with λ the wavelength of light and c its velocity.

3.4.6.1   The External Photo Effect

On the basis of his hypothetical assumptions, Einstein was able to explain the photo effect as outlined above:

When a photon hits a metal surface its energy is captured by an electron according to Eq. (3.103) without any time delay (the probability for an absorption of two photons by a single electron is negligible). By this energy transfer, the electron is liberated, but only then, when $E_{Photon} > \Phi$, i.e. when the energy to release it from the surface is high enough. The resulting kinetic energy, i.e. its velocity, of the electron is the difference between E_{Photon} and Φ.

Figure 3.18 shows a metal plate irradiated by monochromatic light; i.e. all photons possess the same energy. Thus all emitted electrons should have the same energy. By increasing the light intensity, the electron current can now be increased as well.

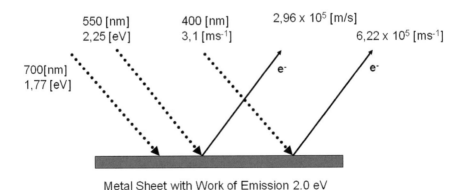

Fig. 3.18 Work of emission and energies

However, by changing the frequency of the incoming light the velocity of the liberated electrons changes as well, confirming Einstein's theory.

At frequencies below the material-specific cut-off frequency ν_0, the electron does not obtain the necessary work of emission, even if the radiation intensity, i.e. the amount of photons reaching the surface per unit time, is increased. Thus the equivalent between work of emission and cut-off frequency can be expressed as follows:

$$h\nu_0 = \Phi \tag{3.104}$$

The electron energy can, by the way, be measured by letting pass the electrons through a break field, which can be generated for example by a spherical capacitor. By varying the tension U in such a field the following dependency is obtained:

$$U \sim E_{kin}/e \tag{3.105}$$

If U becomes higher than the expression on the right side of Eq. (3.105) no current will flow, because the electrons can no longer reach the outer shell. Thus

$$eU = h\nu - \Phi \tag{3.106}$$

Let us imagine the tension figuratively as an inclined plane in space. The inclination of this plane will at some stage be so steep that the electrons can no longer climb up to the end of this ramp. From the gradient of this ramp (the tension measured), one can deduce the velocity of the electrons.

Let us look at the photo effect in more detail: initially, the photon is absorbed by the atomic shell. Its energy is transferred to one electron. Thus this electron either reaches a higher allowed quantum orbit or leaves its formation directly. For that matter, the binding energy of the electron depends on the atomic number of the element as well as on the shell position of the electron. If the photon energy now surpasses the binding energy B of an electron the kinetic energy can be calculated as follows:

$$E_{kin} = h\nu - B \tag{3.107}$$

The scenario described so far concerned the so-called external photo effect, while the raising of electrons onto higher shells by photon absorption is called the internal photo effect (Fig. 3.19).

3.4.6.2 The Internal Photo Effect

Even below the cut-off frequency, physical things happen in the atomic shell. We now look at the case, where the photon energy is insufficient to compensate for the binding energy of an electron. However, the photon frequency in question may still be in a position to lift an electron into an excited state, i.e. onto a higher orbit without

Fig. 3.19 Photo effect

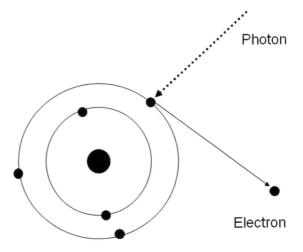

Table 3.4 Works of emission

Element	Φ [eV]	λ_0 [nm]
Li	2.46	504
Na	2.28	543
K	2.25	551
Rb	2.13	582
Cs	1.94	639
Cu	4.48	277
Pt	5.36	231

Source V. Peinhart et al., "Der photoelektrische Effekt", Universität Graz, 2004

liberating it completely from its formation. In metals, we find such excited states in the so-called conduction band, in which electrical current flows. This phenomenon, called the internal photo effect, can be used in photovoltaics or in sensor applications.

Table 3.4 shows works of emission Φ and cut-off wavelengths λ_0 for some metals.

3.4.7 Photovoltaic Cell

The term photovoltaic results from a combination of the Greek word for light and the name of the physicist Alessandro Volta. It denotes the direct transformation of sunlight into electrical energy with the help of photovoltaic or solar cells. The transformation process is based on the photo effect.

Solar cells consist of different semi-conductor materials. Semi-conductors are materials becoming electrically conducting, when light or heat is added, while remaining isolating at deeper temperatures.

More than 95% of all solar cells produced worldwide are made of the semi-conductor material silicon (Si). Silicon has the advantage that there is plenty of supply of it, since it is the second most element in the earth's crust. To make a solar cell work its material has to be "doped". This means the introduction of chemical elements, with which either a positive charge surplus (p-conducting semi-conductor layer) or a negative charge surplus (n-conducting semi-conductor layer) can be obtained in the semi-conductor material.

When two differently doped semi-conductor layers are bound together a so-called p–n junction is obtained. At this junction, an internal electric field is generated leading to a separation of the charge carriers when irradiated by light. Metallic contacts are used to tap the electric tension. When closing the external circuit, i.e. attaching an electric appliance, direct current is flowing.

Silicon cells have the size of about 10 cm × 10 cm. A transparent anti-reflex layer shelters the cell and serves to minimize losses due to reflection at the cell surface.

3.4.7.1 Characteristics of Solar Cells

The tension to be tapped from solar cells depends on the semi-conductor material. For silicon, it amounts to 0.5 V. The disk tension depends only weakly on the incoming light intensity, while current increases with increasing illumination.

The power of a solar cell depends on the temperature. Higher cell temperatures lead to lower power and thus to inferior efficiency. The efficiency tells how much of the incoming amount of light is transformed into usable electric energy.

3.4.7.2 Basic Structure

A classic silicon solar cell consists of an n-layer about 0.001 mm thick inserted into a 0.6 mm thick p-conducting Si substrate. For monocrystalline solar cells, the n-layer is generated by inserting about 10^{19} phosphor atoms/cm^3 into the p-conducting Si substrate close to its surface. The n-layer is thin enough for the sun light to be absorbed in the space-charge region at the p-n junction. The p-conducting Si substrate has to be thick enough to absorb those light rays penetrating deeper and to give it mechanical stability at the same time (Fig. 3.20).

3.4.7.3 Efficiency

The efficiency of solar cells is limited by different loss mechanisms. Basically, the various semi-conductor materials or combinations of them are suited for certain spectral regions of incoming light. A certain part of the radiation energy cannot be

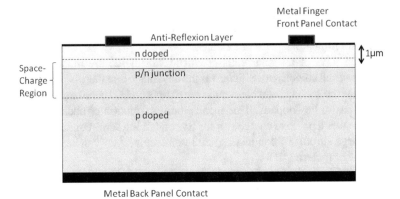

Fig. 3.20 Silicon solar cell

used, because the light quanta to not possess enough energy to activate the charge carriers. On the other hand, a certain portion of photon surplus energy is not transformed into electrical energy but into heat. Furthermore, there are optical losses from shadowing of the cell surface caused by contact pins or reflection of the incoming radiation from the cell surface. Losses through electrical resistance in the semiconductor and connection lines have to be regarded as loss mechanisms. Not to be neglected are also disruptive influences from material contaminations, surface effects and crystal defects. Some of the loss mechanisms (photons with too little energy are not absorbed, the transformation of photon surplus energy into heat) cannot be optimized any further, because they are determined by the material in use. This leads to an efficiency of maximally 28% for crystalline silicon (Table 3.5).

Problem 4: Calculation of Photon Energy

Light with a wavelength of $\lambda = 250$ nm hits a potassium layer with an area $A = 1$ cm². The irradiance amounts to 2 W/m². Calculate the energy of the photon of this radiation.

The Electronvolt

If a particle with the elementary charge $e = 1.602 \ldots \times 10^{-19}$ C passes a potential difference of 1 V (acceleration voltage), its energy acquisition amounts to $1 \text{ eV} = 1.602 \ldots \times 10^{-19}$ J.

Table 3.5 Efficiencies of silicon wafers

Material	Efficiency in % Laboratory	Efficiency in % Operation
Monocrystalline silicon	About 24	14–17
Polycrystalline silicon	About 18	13–15
Amorphous silicon	About 13	5–7

3.4.8 The Compton Effect

The photo effect has shown us that light does not only present itself as a wave but under certain circumstances also as individual quanta, photons, for example in connection with the absorption by an atom. This is proof that electromagnetic radiation has particle qualities as well. Further clear confirmation for this is the Compton Effect, by which an incoming photon liberates an electron from its atomic formation by transferring energy and momentum to it by a scattering process. This experiment as well underlined in its unambiguousness both the particle and wave characters of electromagnetic radiation. Arthur Compton conducted it in 1922.

In this experiment, he directed high-frequency X-rays onto a target of graphite. During his measurement, he detected scattered radiation emanating sideways (Fig. 3.21).

The spectrum of this deflected radiation showed two peaks: one line related to the original incoming wave length, the other had been shifted to longer wavelengths with an increment of $\Delta\lambda$ to the first line. This increment increased, when the angle of observation in this assembly was increased (Fig. 3.22). Just as with the photo effect wave theory could not explain these results.

If again one assumes the particle nature of electromagnetic radiation, the following explanation holds: a light quantum collides with an electron from its atomic shell, transfers energy and momentum onto the latter and is then deflected just as in a classical elastic collision. Expressed in equations:

$$h\nu_E = h\nu + (m_e/2)V^2 \tag{3.108}$$

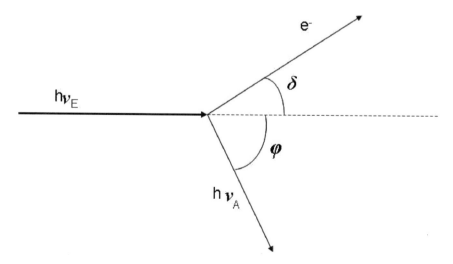

Fig. 3.21 Compton effect

Fig. 3.22 Change of
observation angle

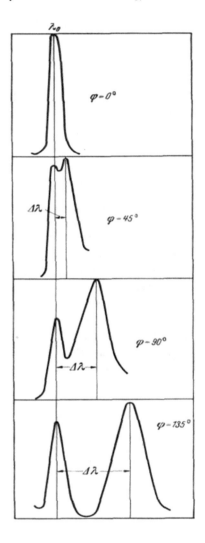

for energy conservation with $h\nu_E$ the input frequency, $h\nu$ the output frequency, m_e the electron mass and v the electron's exit velocity.

$$h\nu_0/c = h\nu \cos\varphi/c + m_e v \cos\delta \qquad (3.109)$$

$$0 = h\nu \sin\varphi/c + m_e v \sin\delta \qquad (3.110)$$

Equations 3.109 and 3.110 represent the conservation of momentum. By conversion and taking into account, that the difference between ν_E and ν is mall, one obtains, in the end, an expression for the difference of the wavelengths:

$$\Delta\lambda = (2h/m_ec)\sin^2\varphi/2 \tag{3.111}$$

h/m_ec has the dimension of a length. It is also called the Compton Wavelength λ_e of an electron. Furthermore $\Delta\lambda$ does not depend on the wavelength of the input frequency but on the angle of deflection φ.

Theoretically, there exists also an "inverted" Compton Effect. In this case, an electron with high energy should meet a low energy photon, transfer its energy and momentum and in this way shift visible light into the X-ray region for example. We can retain as an important result from this experiment as well that light can behave as particles by transferring energy and momentum to another particle.

3.4.9 Material Waves

Another milestone on the path to quantum theory was the discovery that not only light was affected by the duality of waves and particles but matter itself as well. While at the beginning of the twentieth century, people were still convinced that electromagnetic waves, especially light, could be described adequately by their wave characteristics, the photo and Compton effects saw to it that this view of the world was shaken.

Only a few years after the discovery of the Compton Effect, Joseph J. Thomson and other physicists undertook scattering experiments with electrons after some theoretical groundwork by Louis de Broglie. They wanted to find out, whether electron beams would behave similarly to light waves when passing through diffraction gratings. For this, they used thin metal foils. What they observed amounted to a theoretical absurdity at that time. They saw diffraction phenomena at crystals, for example, looking quite similar to diffracted X-rays (Fig. 3.23).

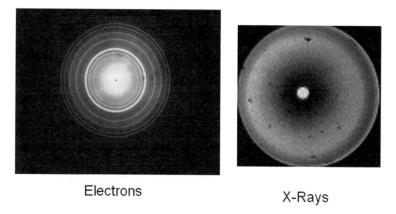

Electrons   X-Rays

Fig. 3.23 Diffraction phenomena (left: Science Source; right: Elsevier)

It seemed as if electrons—like light waves—could annihilate or amplify each other depending on the diffraction angle. Only this was a possible explanation of the characteristic image of light and dark regions on the projection screen. After the experiences with the photo and Compton effects, there existed only one possible explanation for this phenomenon: electrons must also possess wave characteristics besides being a particle. For this, the term material wave was introduced. If particle beams behaved like waves under certain conditions, they must be described by wave characteristics. Thus their wave lengths can be calculated:

$$\lambda = h/p = h/(m * v) \tag{3.112}$$

with p its momentum. If we take for the electron velocity

$$e * U = (m/2) * v^2 \tag{3.113}$$

with U the acceleration voltage, one obtains a relationship between voltage and wavelength. For 10,000 V electrons, their wavelength amounts to $\lambda = 1.2 * 10^{-9}$ [cm]. This corresponds to the wavelength of hard X-rays. Later on, it was discovered that besides electron waves similar ones for atoms or neutrons and other particles could be generated.

These were unequivocal proofs that material particles could also possess wave characteristics. Thus we have the multiple phenomena that electromagnetic radiation, as well as matter, can have both wave and particle characteristics—depending on the experimental setup for their investigation. For both material particles as well as radiation, there exist seemingly no "either-or" but always only an "as-well-as". Thus these phenomena are complementary.

3.4.9.1   Atomic Models

The discovery of radioactivity, the electron as a new atomic particle and many other phenomena incited more and more researchers to occupy themselves with a subject of physics, which had hitherto been regarded as something obscure: the structure of the atom. The earliest models were developed by Thomson and then refined by Rutherford, until the successful combination of Planck's quantum theory with the results of spectral analyses by Niels Bohr led to a sustainable model, which, however, should radically change the conceptualization in classical physical categories.

Let us now look at these models and then relate them to spectroscopic investigations and quantum numbers. After this, we will dive deeper into atomic structures and later get acquainted with two different models of the atomic nucleus. We will discuss the phenomenon of radioactivity, learn about neutrinos and get the first introduction to particle physics.

3.4.9.2 Early Atomic Models

Bohr's elaborations were based on three different preliminary studies:

- the atomic model of Thomson,
- the atomic model of Rutherford,
- the paper by Max Planck about the energy distribution of blackbody radiation.

While earlier on the indivisibility of atoms had been postulated (atoms = gr. indivisible), today this term is used in its original sense for the chemical nature of atoms. From the discoveries at the beginning of the twentieth century, it followed that atoms must have a detailed structure and can no longer be imagined as a compact mass pellet.

Joseph J. Thomson developed a model in 1903, according to which an atom possesses a uniformly distributed positively charged mass, within which also electrons were present. In its ground state, the distribution of the electrons occurred with a minimum of potential energy. Once the electrons were excited they started to oscillate.

The realization of the fact that thin layers of metal were transparent for electrons, served to enhance this model. This was an indication that atoms were not completely filled up with matter, but consisted mainly of empty space. But since they had mass nevertheless, this mass would have to be concentrated in its centre. And there the forces known as electromagnetic would be generated as well.

On the basis of theoretical considerations and experimental results, Ernest Rutherford deduced a diameter of 10^{-12} to 10^{-13} cm for the atomic nucleus and an active radius of the atom itself of about 10^{-8} cm. Since atoms are initially always electrically neutral the number of electrons circling the nucleus had to be equal to the presumed positive atomic number. The atomic number Z in the periodic table of the elements thus corresponds to the number of electrons of an atom.

Rutherford's atomic model assumes that there is an atomic nucleus circulated by electrons at a certain distance. In these models, their centrifugal force neutralizes the attractive Coulomb force between the negatively charged electrons and the positive nuclear charge in such a way that the electrons have a stable orbit and do neither escape from the formation nor fall into the nucleus. However, the main problem of this model is that in this case, we would have to deal with an electric dipole—an oscillating charge, where sometimes the negative charge would appear on one side at other times on the other side of the positive one—continually emitting energy with the result that the atoms would become unstable after a short while. This was a known problem concerning this atomic model.

In the meantime, significant progress concerning the analysis of absorption and emission spectra of atoms and molecules had been made. We will come to the classification of such spectra further down in this chapter. They were the foundation of the atomic model of Niels Bohr.

From the existence of stable atoms, Bohr concluded that at least certain electron orbits must exist, on which electrons could move radiation less contrary to classical electrodynamics (Fig. 3.24). Each of these selected quantum orbits would correspond

Fig. 3.24 Bohr's atomic
model

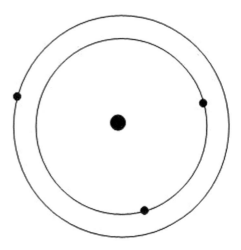

to a certain energy state E. For that matter, the quantum orbit with the smallest radius
would correspond to the ground state of an atom. If one wanted to transfer an electron
to an orbit further away from the nucleus excitation energy would have to be added.
After this, the electron would jump back after an extremely short time (10^{-8} s) onto
an energetically lower orbit. In the process energy would be released and emitted as
a spectral line with a frequency ν:

$$E_a - E_c = h * \nu \tag{3.114}$$

with E_a the energy of the higher orbit and E_c the energy of the lower one. The energy
packet liberated (photon) with the energy $E = h*\nu$ can then be registered and its
characteristics for each orbital transition. Bohr developed the following criterion for
permitted quantum orbits:

$$2\pi \, r \, m \, v = n \, h \text{ mit } n = 1, 2, 3, \ldots \tag{3.115}$$

It is obvious that an ever-increasing excitation of an atomic electron would finally
lead to its complete separation from the atom, so that this cut-off energy has to be
equal to the ionization energy of the atom.

Definition 1 An ion is a negatively or positively charged atom or molecule (a chemical compound from two or more atoms) either lacking electrons or having a surplus
of electrons.

Bohr's atomic model was later replaced by the quantum mechanics of Schrodinger
and Heisenberg, but had been essential for these further developments.

3.4.9.3 Spectra

As already mentioned, findings from spectroscopy have significantly influenced atomic model theory. We have to differentiate between absorption and emission spectra. Their visible difference is the occurrence of bright bands or lines on dark background for absorption spectra (Fig. 3.25), while emission spectra appear dark on a bright background. Absorption spectra from white light comprising many frequencies are seen, when light passes through a gas to an observer for example. Some of those frequencies are absorbed by the gas and thereafter re-emitted into all directions. The light received by an observer is bright for all frequencies except for those absorbed and scattered around. These frequencies are recorded as black bands or voids in a spectrum. They are characteristic of different gases. Emission spectra are created, when materials emit specific frequencies without prior irradiation from a different light source. These frequencies are then registered as bright bands. Hot bodies for example possess characteristic emission spectra.

Finally, one differentiates between line, band and continuous spectra. Line spectra are always emitted or absorbed by atoms, while molecules (many atoms) are responsible for band spectra.

For a better understanding, let us consider a relatively simple spectrum—that of the hydrogen atom. In its simplest occurrence, the nucleus of hydrogen is a single proton surrounded by a shell made up of one electron.

When analysing the hydrogen spectrum one realizes that apparently, it does not follow an arbitrary pattern but specific laws. It is obvious that the spectrum obeys a series. It was Johann Jacob Ballmer who first investigated such a series, the first four lines of which were later named after him: the Ballmer Series (Figs. 3.26 and 3.27).

There exist specific series for all elements. To better classify them the wave number was introduced:

$$\bar{v} = 1/\lambda \left[\text{cm}^{-1} \right] \tag{3.116}$$

Johannes Rydberg developed a general formula to calculate line series on this basis:

$$\bar{v} = R/(m + a)^2 - R/(n + b)^2 \quad \text{mit } n > m \tag{3.117}$$

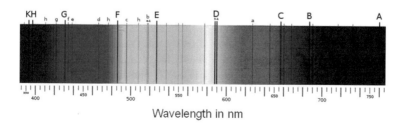

Wavelength in nm

Fig. 3.25 Absorption spectrum

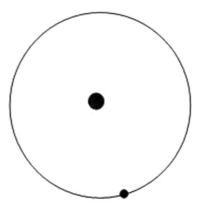

Fig. 3.26 Hydrogen atom

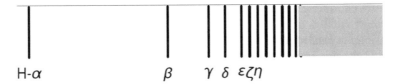

H-α   β   γ δ εζη

Fig. 3.27 Balmer series for hydrogen

with R the so-called Rydberg Constant; a and b are specific constants of a particular series, m denotes a specific series in a spectrum, and n is a running number. The Ballmer Series of hydrogen, for example, can thus be expressed the following way:

$$\bar{v} = R/2^2 - R/n^2 \text{ mit } n = 3, 4, 5, \ldots \tag{3.118}$$

The Rydberg Constant for Hydrogen is $R_H = 109{,}677.576$ cm^{-1}. The in-depth explanation for such a series is rather ambitious and will not be detailed here.

3.4.10 Quantum Numbers

If we now bring together the results of spectral analysis and Bohr's model, we first encounter two conditions, which follow directly from that atomic model. One is a quantization condition, according to which only for certain whole numbers n combinations of the electron's distance (radius r) to the nucleus and its velocity (v) do exist:

$$2\pi * r * m * v = n * h \quad \text{with } n = 1, 2, 3 \ldots \tag{3.119}$$

The other is found in the formula for the angular velocity of electrons depending on the radius:

$$e^2/r^2 = m * r * \omega \tag{3.120}$$

with ω the angular velocity of an electron, where in this case the centrifugal force corresponds to the Coulomb attraction. From this we then obtain the electron orbits $r_1 \ldots r_n$:

$$r_n = h^2 * n^2 / (4\pi^2 * m * e^2) \tag{3.121}$$

as well as

$$\omega_n = 8\pi^3 * m * e^4 / (h^3 n^3) \tag{3.122}$$

for each n. n is called a quantum number. One consequence among others from this is that angular momentums $mr^2 \omega$ do only exist as whole number multiples of $h/(2\pi)$. This is in glaring contradiction to classical mechanics, where the angular momentum is a continuous quantity, and planets can circle the sun theoretically on all possible orbits and not only on some preferred ones. Furthermore, r_n is that particular radius corresponding to the n-th quantum orbit.

If we want to calculate the energy level of a quantum orbit, we have to take into account the total, i.e. the potential and kinetic energy of a quantized state:

$$E_n = (1/2)I_n\omega_n^2 - e^2/r_n \tag{3.123}$$

with I_n the so-called moment of inertia of the atom in the state n with $I = mr^2$. Then we obtain the solutions for r_n and ω_n:

$$E_n = -2\pi^2 m e^4 / h^2 n^2 \quad \text{with } n = 1, 2, 3, \ldots \tag{3.124}$$

Since n is in the denominator the energy decreases at higher orbits (a greater n in the denominator leads to a smaller value in the overall expression). But since the whole expression is preceded by a negative sign, greater (negative) values mean smaller values—in the sense that -1000 is smaller than -1. In this sense, the energy of the innermost orbit is the lowest—and increases with increasing distance against zero.

Figure 3.28 summarizes once more the most important series of the H-atom with their energy levels.

n is called the main quantum number. Our considerations up to now were based on the simplest atom, hydrogen. If one proceeds from there along the periodic table of the elements the atomic composition gains in complexity: there exist several electron shells nested into each other influencing each other mutually with consequently more complex spectral lines. This has necessitated the introduction of further quantum numbers: l stands for the orbital angular momentum and s for the spin of the electron;

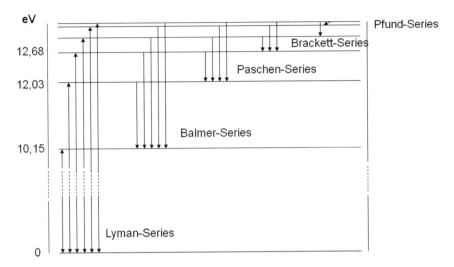

Fig. 3.28 Excitation series of hydrogen

their vector sum adds up to:

$$j = l + s \tag{3.125}$$

for the total angular momentum. Without going into the exact derivation, a principle is announced here, the Pauli Principle (after Wolfgang Pauli), which organizes the relationship between these four quantum numbers n, l, s and j:

Definition 2 "Only such electron combinations in atoms and molecules exist in nature, in which all electrons differ with respect to at least one of their four quantum numbers."

The Pauli Principle sees to it that only one electron may occupy any possible position in an atom, and all other electrons entering an atom are supplanted from such a position. In very dense stars, so-called neutron stars, for example, this repulsion opposes the gravitational force; the Pauli repulsion between electrons on atomic orbits prevents the implosion of the star due to gravitational pressure.

After the introduction of further quantum numbers in elementary particle physics this principle was generalized for all of quantum physics.

3.5 Nuclear Physics

3.5.1 The Periodic Table of the Elements

The periodic table of the elements (Fig. 3.31) had initially been a subject of chemistry and later of atomic physics, but led later to general considerations concerning nuclear theory as the starting point for the analysis of mass and structure of atomic nuclei (Fig. 3.29).

The classification concerning the structure of the electron shell of each atom by adding one more electron to a preceding element had been developed from spectral analysis and Bohr's model with its quantum numbers. In this model, the atomic shell itself consists of electron shells manifesting themselves as periods. Each element is labeled with a number, the atomic number, corresponding to the number of electrons and thus protons in the nucleus (atoms are neutral; therefore, the last two numbers are equivalent). Of interest, however, in this context is not so much the atomic number but the weight of each atom.

If one now imagines that all atoms are constructed from hydrogen, one will find out very quickly, that they are much heavier than one would suppose on the basis of

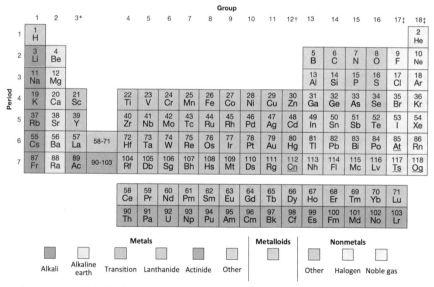

*The placement and display of lanthanides and actinides is not the same on all periodic tables due to differing scientific views on the composition of Group 3. The last two members of the group are also known as transition metals.

†Elements Zn, Cd, and Hg are sometimes considered as *other metals* rather than *transition metals*.

‡The *halogen* and *noble gas* descriptors are commonly applied to all members of their respective groups.

Predictions for underlined elements: 85-At and 117-Ts are metals; 112-Cn is a non-metallic noble liquid; 118-Og is a metallic-looking reactive solid nonmetal. These predictions await experimental confirmation.

Fig. 3.29 Periodic table of the elements (Sandh, CC BY-SA 4.0 https://creativecommons.org/lic enses/by-sa/4.0, via Wikimedia)

their atomic number and on top of that—assuming H has an atomic weight of 1—that they are not whole number multiples of the weight of hydrogen. We know that atomic nuclei do not only consist of protons but also neutrons. Initially, people have tried to derive the non-integrity of the atomic weights from the difference between the real mass of the proton and the neutron.

But there is another way to resolve the puzzle by using a different standard than H, for example, ^{12}C (the type of carbon most common in nature, consisting of 6 protons and 6 neutrons), and fixing its reference mass at 12. In this way, however, one obtains a value of 1.008 for H instead of 1. The only possible explanation for the deviations from integrity was based on the fact that there is no obvious correlation between atomic number and atomic weight. This means that one specific atomic number may represent several different nuclei of the same element. These will always possess the same positive charge according to their atomic number but in addition in each case a different number of neutrons.

Definition 3 Atoms of equal charge but with a different number of nucleons (nucleon is the collective term for protons and neutrons as components of the atomic nucleus) are called isotopes of an element.

Therefore the atomic weight indicated in the periodic table is always the weighted average of all isotopes of the element in question. The weight is derived from the fraction of an isotope found in nature as measured by the sum of all isotope fractions of the same element. Normally, no one conducts any isotope separation before weighing an element. Thus there is always a naturally occurring isotope mix in hand.

3.5.2 The Strong Interaction

Because the negative charge of the electrons is compensated by the sum of the positive charges of the protons in the nucleus electrical neutrality of an atom is ensured. But what force keeps the positively charged protons together, since they should repel each other because of their electromagnetic charge? The reason cannot be the gravitational forces exerted by the neutrons because of their vanishingly small effect. Therefore, the nuclear building blocks, the nucleons, must be bound by much stronger forces. This means that besides gravitational and electromagnetic interaction an additional natural force, the strong interaction, must exist.

This strong interaction is the strongest force in nature in comparison, but has only a small reach (Table 3.6).

Table 3.6 Strength and reach of forces in nature

Force	Strength	Reach
Gravitation	10^{-39}	∞
Electromagnetism	10^{-2}	∞
Strong interaction	1	10^{-15} [m]

3.5.3 Structure of the Atomic Nucleus

We have already learned that the atomic nucleus is made up of protons and neutrons, which are slightly heavier than protons. An isotope is characterized by two parameters:

- the atomic number Z (number of protons),
- the mass number A (number of nucleons, i.e. neutrons plus protons).

Hence the number of neutrons in an isotope nucleus can be calculated as follows:

$$N = A - Z \qquad (3.126)$$

The atomic nucleus, the diameter of which can never be determined exactly for fundamental reasons but has an order of magnitude of 10^{-12} cm, contains nearly the total mass of an atom with the exception of the electrons contributing, however, only an unimportant part to the total mass. The resulting density of the nucleus amounts therefore to 10^{14} g/cm^3 or—to express it differently—a mass of about 100 million t is compressed into one single cm^3 of nuclear matter.

Just as electrons protons and neutrons possess a spin of $\hbar/2$ as well with

$$\hbar = h/(2\pi) \qquad (3.127)$$

This definition is called the quantum of angular momentum. Just as its nucleons, the atomic nucleus itself has a spin:

$$|I| = I * \hbar \qquad (3.128)$$

with I the nuclear angular momentum quantum number. |I| is the vector sum of its own angular momentum and the spins of the protons and neutrons constituting the nucleus. The total angular momentum J is calculated in analogy to the one for the electron shell. Since the spin of the nucleons is half-integral, nuclei with an even mass number have an integral angular momentum (quite often $= 0$), and those with an odd mass number are always a half-integral one. Furthermore, all nuclei possess a magnetic momentum together with its total angular momentum similar to the electron shell.

3.5.4 Nuclear Models

Even today, we do not have a consistent model for the atomic nucleus, which could describe all its features. All theoretical models do present certain phenomena approximately correctly, but have to be adjusted or replaced by alternatives for other phenomena.

Similar to the model of the atom, a shell model of the atomic nucleus has been developed, for example, in which the arrangement of the nucleons and their systematics has been projected. But this model only explains in part and approximately certain observations, for example, nuclear energy spectra. Other differences in comparison to the atomic shell can be found in the combination of the quantum numbers and the angular momentums. In the following, we will limit ourselves to two nuclear models, which play important roles in experimental practice:

- the liquid drop model,
- the optical model.

3.5.4.1 The Liquid Drop Model

Let us begin with the liquid drop model. It had been developed in the thirties of the twentieth century by Carl Friedrich v. Weizsaecker and looks at the atomic nucleus as if it were a liquid drop. Such a consideration makes sense because of the constant density of all atomic nuclei and—as we shall see later—the constant binding energy per nucleon with the exception of very light nuclei. One result of this model is that the binding forces within the nucleus are only acting between directly adjacent nucleons. Binding energy means in analogy to the electron shell in atomic physics those energy levels, which bind a nucleon to the nucleus. It is measured in [MeV]—mega electron volt.

On the basis of the liquid drop model v. Weizsaecker developed an equation, which expresses the binding energy per nucleon as a function of the mass number of the nucleus. For this he introduced five terms:

1. a term a_1 for the mean binding energy of a roundly bound nucleon;
2. a term taking into account that, while neglecting Coulomb repulsion between protons, the binding strength is at a maximum for equal proton and neutron numbers (derived from the shell model):

$$a_2 * ((N - Z)/(N + Z))^2 \qquad (3.129)$$

3. a term taking into account the surface tension in such a way that for the outer most neutrons there is only one-sided binding from inside:

$$a_3 * (N + Z)^{-1/3} \qquad (3.130)$$

4. a term taking into account electrostatic repulsion:

$$a_4 * Z^2/(N+Z)^{4/3} \tag{3.131}$$

5. a term allowing for the fact that nuclei with even proton and neutron numbers do possess a somewhat higher binding, nuclei with double uneven nucleon numbers a somewhat smaller binding relative to the combination even-uneven. The reason for this is to be found in the spin combination of the nucleons:

$$a_5 * (N+Z)^{-2} \tag{3.132}$$

The terms a_1, a_2, a_3, a_4 and a_5 can at present not be derived theoretically and are based on empirical determinations. If we now put everything together, we obtain for the binding energy per nucleon the equation named after its creator, the Weizsaecker equation:

$$E\,[MeV] = 14.0 - 19.3\big((N-Z)/(N+Z)^2 - 13.1/(N+Z)^{1/3} - 0.60$$
$$* Z^2/(N+Z)^{4/3} \pm 130/(N+Z)^2 \tag{3.133}$$

It is rather astonishing, when contemplating this equation, that it does completely without any complicated mathematics like for example partial differential equations for such a complex task as developing an atomic model.

To be able to interpret the results of the Weizsaecker graph (Fig. 3.32), we have to introduce the term *mass defect*. Because measurements have shown that, when adding all particles of a nucleus together, the total mass sum would have to be greater than the actual measured total mass of the same nucleus. At first sight, it seemed that this result contradicts the law of the conservation of mass as known from chemistry. But we know from the theory of relativity that mass and energy are equivalent. The observed mass defect thus corresponds to the binding energy of the nucleons (Fig. 3.30).

The shape of the graph shows that the binding energy per nucleon starts to diminish after a maximum at mass numbers from about 50–70 for increasingly heavy nuclei. During nuclear fission, a heavy nucleus at the right end of the graph is split into two nuclei with mass numbers between 80 and 160. In this split, the original mass defect is responsible for the enormous liberation of binding energy. On the extreme left-hand part, we observe the light nuclei, the binding energy difference of which is still much greater, when, for example, two light nuclei fusion. For this reason, during nuclear fusion, a very much higher energy potential is liberated in comparison to nuclear fission. The energy radiated from our sun originates from these nuclear fusion events.

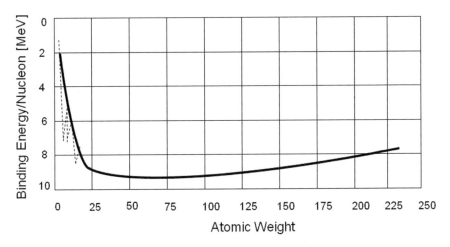

Fig. 3.30 Binding energy per nucleon

3.5.4.2   The Optical Model

The liquid drop model describes certain aspects of nuclear fission very well, but it is less useful for other nuclear reactions such as, for example, absorption or scattering. One model, which is more useful for these or others, is the optical one.

Let us take the case that an external neutron enters an atomic nucleus. Two alternatives are possible:

- The neutron moves independently about the atomic nucleus and leaves it immediately.
- The neutron gets absorbed from the nucleus and constitutes a different nucleus together with the target nucleus.

Both cases concern either absorption or re-emission. To describe both reactions, the optical model introduces a mathematical complex optical potential acting on a neutron like a lens does on light rays:

$$U = V(r) + iW(r) \tag{3.134}$$

with V as a potential sink comprising all relevant energy states. W is responsible for the absorption effect. r is any space coordinate. This potential can now be introduced into Schrodinger's equation and will finally—after passing some longish mathematic path—obtain probabilities for

- elastic scattering,
- absorption,
- the overall reaction.

With the help of the optical model, transmission coefficients can be determined quantifying the penetrability of a nucleus. In the overall context of all nuclear reactions conclusions about reaction probabilities are obtained. The latter are called cross-sections in physics. Their unity is [barn] with 1 [barn] = 10^{-24} [cm^2].

3.5.5 Radioactivity

3.5.5.1 History

The Curies

In the summer of 1903, Ernest Rutherford and his wife Mary visited the Curies in Paris. By chance, Marie Curie was rewarded her Ph.D. just the day the Rutherfords arrived. Mutual friends had arranged a celebration party. "After a rather lively evening", Rutherford remembered," we retired into the garden at about 11 p.m., where Professor Curie shows us a tube, which was partly covered with zinc sulfide, and which contained a fair amount of radium in a solution. Its luminosity was brilliant in the darkness, and all this presented a fitting finale of an unforgettable day." The zinc sulphide cover fluoresced white and took its way along the periodic table from uranium until lead, visible in the darkness of this Paris evening by emitting high-energy particles. The light was bright enough to enable Rutherford to see the hands of Pierre Curie, which were "in a very inflamed and painful state, after having been exposed to the radium rays in such a way." Swollen hands from radiation burn were further confirmation of the energy potential contained in matter.

While at that time a young woman was supposed to keep her household in order, care for a baby and place pans on an oven, a female physicist made one of the most important discoveries of modern science in a run-down laboratory at the School of Physics in Paris.

After Roentgen's discovery of X-rays, Henri Becquerel wondered about the possibility, whether one could find out, if these rays were perhaps being emitted from fluorescent bodies after a prior irradiation by light.

Henri Becquerel was the first physicist, who observed a phenomenon, which was later labeled "radioactivity" by Marie Curie. But the nature of this radiation remained a mystery for some time.

Marie Curie had to find a laboratory, where she could carry out her experiments with pitchblende. She was provided with a small studio behind glass walls at the ground floor in the School of Physics, after her husband had consulted with the director. This place was something like a damp store room filled with useless machines and wood. The technical facilities were rudimentary; the comfort was equal to zero. Her technical equipment consisted of an ionization chamber, an electrometer and a piezo-electric piece of quartz.

After several weeks first results became visible. Marie Curie had the impression that this mysterious radiation was of an atomic nature. Perhaps it had just been an accident that this radiation had been discovered by Becquerel initially in connection with uranium and that for this reason it would always be associated with uranium. Now she had to look elsewhere as well. Why should other materials not possess the same characteristics?

She proposed the term "radioactivity" as a denomination. But the discovered radioactivity was significantly higher than guessed from the amount of uranium and thorium processed.

After ten or twenty repetitions, she remained with her conclusion that the uranium and the thorium in the experimental charge could not alone be responsible for the intensity of the radiation.

The scientist answered this question with certainty: these minerals must have contained some radioactive substance being at the same time a hitherto unknown chemical element—a new element.

Only four years ago she had written: "Life is not easy for us. But, who cares? We have to have perseverance and trust in ourselves on top. We have to believe that we have some sort of vocation, and that we have to strive for it, whatever the costs."

In a first communication to the Academy, presented by Prof. Gabriel Lippmann and published in its Proceedings on 12 April 1898, Marie Sklodovska Curie announced the possible presence of a new element with strong radioactivity in pitchblende ores. This was the first step in the discovery of radium.

The vigour available had now been doubled after her husband participated in the experiments. Two brains, four hands were now looking for the unknown element in that small damp workshop in Rue Lomond. From this instance onwards, it became impossible to distinguish the different contributions to their common endeavor of each of the two.

In July 1898, they were in a position to announce the discovery of this substance with certainty. In the Proceedings of the Academy one could read:

"We believe that this substance, which we extracted from pitchblende, contains a metal, which has until now not been observed, and behave like bismuth concerning its analytical properties. If the existence of this new metal will be confirmed, we propose to call it Polonium, derived from the name of that country, which one of us comes from.

They lived in a small flat in Rue de la Glacière. The paper had been drafted by Marie and Pierre Curie and a collaborator named Gustave Bémont. Here are some lines from it concerning Radium: "All reasons cited by us have convinced us that the new radioactive substance contains a new element, for which we propose the name RADIUM. This new radioactive substance contains with certainty an important portion of barium; in spite of that its radioactivity is enormous. Therefore, the radioactivity of radium has to be enormous as well." […].

Becquerel

A few months after Roentgen had published his discovery of X-rays, Becquerel made the discovery of radioactivity, which proved equally sensational for the world afterwards. His findings indicated that atoms must have a structure, although at that time no one was capable to interpret it. Indeed, Becquerel's discovery did not make a big impression since it had been made virtually within the context of Roentgen's work. He was led to his discovery after testing Roentgen's observation that X-rays were deflected from the glass wall of a tube after bombarding it with cathode rays. Furthermore, the glass showed visible fluorescence at these points. Assuming a connection between this fluorescence and the emission of X-rays Becquerel made up his mind to investigate different materials known to fluoresce, among them certain uranium compounds. After a number of experiments, Becquerel concluded that those uranium compounds, which he had investigated, emitted rays, which could penetrate materials of different thicknesses.

Initially, it was assumed that these were some kind of X-rays. There were a number of similarities: the power of penetration of the radiation, the fact that photographic plates could be exposed, and last but not least the observation that the radiation in question ionized air during its passage. The latter was especially important since it presented a simple method to discover radiation and measure it. Shortly after Becquerel's observations, it was found that thorium also possessed similar qualities, and several years later Pierre and Marie Curie finally discovered other radioactive substances, among them radium.

Antoine Henri Becquerel was born on 15 December 1852 in Paris—the third in a row of important scientists. His father as well as his grandfather had been renowned physicists. Each of them occupied the chair of physics at the Musée d'Histoire Naturelle in succession, and each influenced the choice of the young Becquerel's career in his particular way. His contemporaries were Hertz, Roentgen and J. J. Thomson, all of whom played important roles in the early stages of quantum physics—that epoch, which would become the most fruitful period in the history of science. After quite ordinary schooldays Becquerel visited the École Polytechnique, where he obtained his Ph.D. In 1875 he took up an appointment to serve the government as an engineer in the department of roads and bridges. In 1894, he became chief engineer. In the meantime, he taught physics, where his father and grandfather had done the same, and in 1882, after his father died, he followed him in his chair. In 1889 he was elected member of the Institute de France and in 1895 he became professor for physics at the École Polytechnique.

His early endeavors dealt basically with different properties of light including polarization, absorption in crystals and phosphorescence. In 1896, he discovered radioactivity, initially called Becquerel rays. Thereafter he continued his investigations in this field until his death in Brittany in 1908. He was awarded the Rumford Medal of the Royal Society as well as the Nobel Prize together with Pierre and Marie Curie in 1903.

The starting point for Becquerel's investigations was the discovery that photo plates, which he needed for experiments on phosphorescence, had already been exposed when he wanted to employ them. They had been stored close to uranium salts, which he wanted to investigate. After this chance discovery, he proceeded systematically to get to the bottom of this phenomenon.

He carried out the following experiments one after another:

(1) A photo plate with a bromide emulsion was wrapped in two layers of thick black paper. The paper was thick enough to prevent any exposure over a complete day. Then he placed a piece of a phosphorescent substance on top of it and left everything to daylight for a whole day.

(2) Thereafter he placed a coin and a piece of sheet metal of a distinct shape between the photo plate and the uranium compound.

(3) Next he covered the plate with an aluminium sheet and placed uranium salt on top of it.

(4) As a next step he inserted a piece of copper sheet between the aluminum and the photo plate.

(5) To exclude chemical reactions, he prepared the uranium salt in a convex shape to touch the photo plate only at one single point.

(6) Eventually he constructed a mount, with which the salt could be suspended above the photo plate.

In the simplest case, a dark spot appeared on the photo plate (Fig. 3.31).

Concerning the coin and the metal sheet (a Maltese cross), the contours of both objects became visible in front of a bright background.

The intensity of the contrast level of the photo emulsion changed with the thickness of the aluminium sheet. The copper sheet was indeed able to attenuate the radiation but not stop it completely.

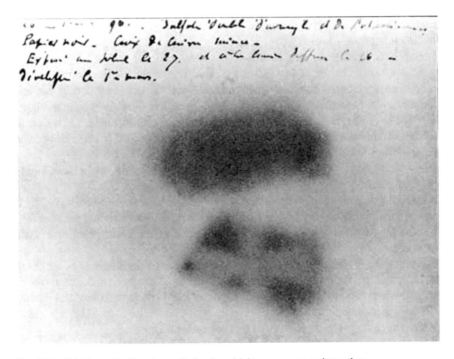

Fig. 3.31 Shielding of radioactive radiation by a Maltese cross on a photo plate

After having excluded the phosphorescence of the uranium salt as well as sun light as sources for these phenomena, he concluded that it was the uranium compound emitting some sort of radiation capable of penetrating thick paper, aluminium and copper. Later he employed different uranium compounds and found that the radiation did not originate from the phosphorescent salts but from the uranium itself. At that time, Becquerel was not able to give any explanation of these observations.

3.5.5.2 Physics Background

Radioactivity had been discovered at the beginning of 1896, but Ernest Rutherford identified this radiation only three years later. He found out that it consisted of two components:

- one, which was absorbed rather promptly by thin layers,
- another one, which was much more penetrating.

He called them alpha- and beta-rays. Later it was found that yet another even more penetrating radiation, gamma-radiation, was accompanying these two other types of radiation.

In nuclear decay, three types of radiation are thus known (Fig. 3.32):

α-decay,
β-decay,
γ-radiation.

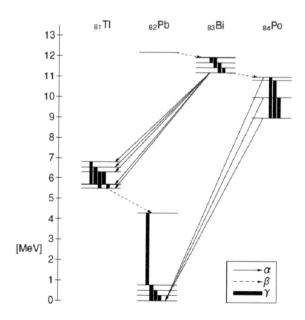

Fig. 3.32 Radioactive series until a stable lead isotope

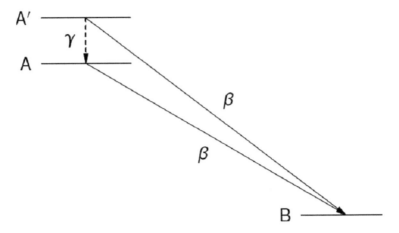

Fig. 3.33 Isomeric nuclear decay

In α-decay, the nuclear charge, i.e. the atomic number of an element, decreases by two units. α-rays consist of two protons and two neutrons, are thus positively charged. An α-particle corresponds to the nucleus of the helium atom. In β-decay the atomic number is increased by 1. β-rays are negatively charged and are electrons. γ-radiation is of electromagnetic nature and neutral—basically high energy light. It does not influence the atomic number. γ-radiation accompanies both α- or β-decay.

Radioactive nuclei have a probability of decay. This depends on the number of nuclei still present in a material. The age of a specific nucleus is unimportant for it. The probability of decay is measured in half-life.

Definition 4 The half-life is the time span, after which half of the initially present number of nuclei of a certain type has decayed.

Radioactive nuclei can have half-lives between 10^{-7} [s] and more than 10^{14} years.

Since γ-radiation is electromagnetic it shows similar characteristics as visible light: its spectra indicate excitation states of the atomic nucleus quite similar to energy levels in the electron shell in atomic physics (Fig. 3.33).

It has been observed that β-particles do not indicate discrete energy states but follow a continuous spectrum. This data is difficult to reconcile with the fact that the original nucleus as well as the final one are very well found to be in discrete energy states. Furthermore, there are no electrons in a nucleus. These two problems are resolved by the following relationship (Fig. 3.34):

$$n \rightarrow p + e^- + \bar{\nu}_e \tag{3.135}$$

Relationship (3.135) expresses the transformation of a neutron into a proton. In this decay, a free electron is generated. This explains the upward change of the atomic number on the one hand and on the other the origin of electrons. $\bar{\nu}_e$ is a new particle, called the neutrino, which is set free during decay. Neutrinos possess only

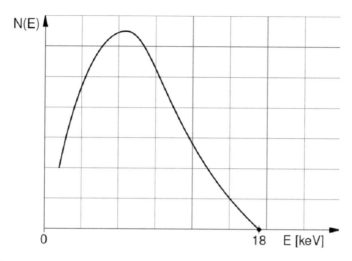

Fig. 3.34 β-spectrum

an extremely tiny mass. For reasons of energy conservation, its existence is required for the reaction described above. But there are two more reasons for the existence of neutrinos: die laws of conservation of momentum and angular momentum of an atomic nucleus. To be more precise: during the transformation of a proton it is not a neutrino, which is set free, but an anti-neutrino, expressed by the bar above the neutrino symbol.

Weak interaction is responsible for β-decay. In it, a neutron changes into a proton, and an electron and the corresponding anti-electron-neutrino are generated. Now we know that nucleons are not elementary but consist of quarks. Therefore in β-decay, an up-quark has to become a down-quark as depicted in Fig. 3.35.

A neutron has two down-quarks and one up-quark, a proton two up- and one down-quark. The transformation of a down- into an up-quark happens under the emission of an electron and an anti-neutrino. In this way, the charge balance is preserved. The dotted lines in the figure between the quarks symbolize the binding by gluons; the "sea" of further virtual gluons and anti-quark pairs is left out in the picture for reasons of clarity.

Meanwhile, we have noticed a nearly uncountable artificial radioactivity besides natural radioactivity, which has been generated and is still going on to be generated by human intervention like bombarding atomic nuclei and nuclear fission.

For a nucleus to be stable at all a configuration of nucleons must be able to check the attractive interaction of the nuclear forces against the repulsive Coulomb forces. Nuclei, which have a ratio of the number of neutrons to the number of protons of between 1 and 1.56, and for which the number of protons does not exceed 82, do fulfill this condition. Nuclei outside this "valley of stability" attempt to reach this range of stable nucleon configurations by emitting radiation. One distinguishes such unstable nuclei found in nature—so-called natural emitters—from such nuclei, which

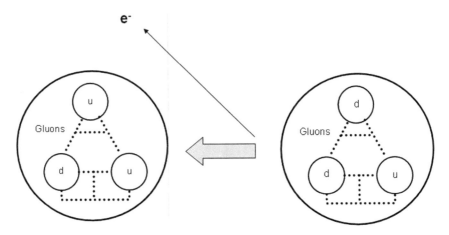

Fig. 3.35 β-decay

are produced from originally stable isotopes by adding energy or bombarding them with other nuclear particles—the already mentioned artificial radio isotopes. The same laws apply independently from those formal differences.

Radioactive transformations or decays of unstable atomic nuclei are distinguished by three main characteristics:

- the type of the particle emitted from a nucleus,
- the energy of that particle,
- the time span necessary for the decay of half the number of originally present atomic nuclei, the so-called half-life.

Type and energy of the emitted particle depend on the nucleonic configuration and the state of excitation of the radioactive atomic nucleus. As already mentioned, α-particles are twice positively charged He nuclei. They consist of two protons and two neutrons, representing a particular stable configuration. They are emitted preferentially by heavy unstable nuclei ($Z > 82$). If we designate an unstable nucleus quite generally by X and the successor nucleus after emission of a particle by Y, α-decay looks like:

$$_{Z}X^{A} \rightarrow {}_{Z-2}Y^{A-4} + {}_{2}\alpha^{4} \tag{3.136}$$

This general equation can be illustrated by the α-decay of the uranium isotope $_{92}U^{238}$:

$$_{92}U^{238} \rightarrow {}_{90}Th^{234} + {}_{2}\alpha^{4} \tag{3.137}$$

The original nucleus $_{92}U^{238}$ decays from its ground state to the nucleus $_{90}Th^{234}$, a thorium isotope, which in turn is also unstable. The energy of the emitted α-particles appears at two different levels: $E_1 = 4.18$ meV at about 77% and $E_2 = 4.13$ meV at about 23%. E_1 corresponds to the transition to the ground state of $_{90}\text{Th}^{234}$, while E_2 must be related to the transition to an excited state of $_{90}\text{Th}^{234}$. This excited energy state of about 50 keV descends to the ground state after the emission of a γ-quant. This is an example to illustrate that during radioactive decay of one type of nuclide in many cases not only one but several particles, in this case, α and γ can be emitted, which differ both in type and energy.

The β-particles emitted from unstable nuclei are negative or positive electrons, generally called β^- and β^+, when they originate from the atomic nucleus. The emission of a β^--particle, β^--decay, happens, when the unstable atomic nucleus tries to enter into a stable nucleon configuration by transforming a nuclear neutron into a nuclear proton. Hence β^- decay can be found for nuclei with neutron surplus. Unstable nuclei with a proton surplus look for stability by transforming a nuclear proton into a nuclear neutron by emitting a β^+-particle. Hence in β^--decay the atomic number Z for the successor nucleus increases by 1, while it decreases by 1 in β^+-decay. The mass number A remains constant in both types of decay. The general expressions are

$$\beta^--\text{decay:} \quad _z X^A \rightarrow {}_{z+1} Y^A + \beta^- \tag{3.138}$$

$$\beta^{-+}-\text{decay:} \quad _z X^A \rightarrow {}_{z-1} Y^A + \beta^+ \tag{3.139}$$

As an example for β^--decay let us look at the already introduced $_{90}\text{Th}^{234}$:

$$_{90}\text{Th}^{234} \rightarrow {}_{91}\text{Pd}^{234} + \beta^- \tag{3.140}$$

The Thorium isotope decays to Protactinium, which is again unstable and continues to decay by emitting a β-particle. In these β-decays transitions to excited states of the successor nucleus may happen, so that the emission of γ-Quanta takes place at the same time. The negative electron emitted during the decay remains stable as a particle. The same is true for the positron emitted during β^+-decay; however, at the end of its trajectory, it unites with a negative electron and both annihilate each other into 2 γ-Quanta, termed as annihilation radiation.

There exists a further method to change a nuclear proton into a nuclear neutron by capturing an electron from the electron shell into the nucleus. After this so-called electron capture, the atomic number of the successor nucleus decreases by 1 just as in β^+-decay. The shell electrons can be captured from any of the occupied electron shells, but the capture from the shell closest to the nucleus is the most probable. When the liberated orbit is replaced by other shell electrons distinctive X-rays are emitted by the successor nucleus, so that this process can be identified.

3.5.5.3 The Law of Decay

Let us assume that the mass of a radioactive substance consists initially of N atoms. The next question is: how many decays will happen per unit time? Or to put it differently: how important is the radioactive "activity" of this isotope? What do the underlying laws responsible for diminishing activity look like?

To answer these questions, we have to introduce the radioactive decay constant λ. It tells us something about the probability of the decay of any nucleus. The activity is then:

$$A = \lambda N \tag{3.141}$$

On this basis, we get a reduction of the number of radioactive nuclei over time:

$$dN/dt = -\lambda N \tag{3.142}$$

The total number of decays over a time interval amounts to

$$\int_0^t dN/N = -\int_0^t \lambda dt \tag{3.143}$$

If N_0 is the number of radioactive nuclei at the time $t = 0$, we obtain

$$\ln(N/N0) = -\lambda t \tag{3.144}$$

with
  N = the number of nuclei at the time t,
  N_0 = the number of nuclei at the time $t = 0$,
  λ = decay constant.

Equation (3.144) is called decay law. The decay constant, which is specific for each isotope and thus for radioactive decay itself, cannot be influenced neither physically nor chemically. Decay follows statistical laws. But because of this fundamental nature more precise assertions can be deduced by observing more and more decay events.

Henceforth there is the possibility that one and the same radio isotope may decay in several different ways. In this case, the total decay constant consists of the sum of all probabilities for different types of decay:

$$\lambda = \lambda_1 + \lambda_2 + \lambda_3 + \cdots \tag{3.145}$$

And for the partial activity:

$$dN_i/dt = -\lambda_i N = \lambda_i N_0 e^{-\lambda_t} \tag{3.146}$$

with:

dN_i/dt decays of decay type i.
N   number of nuclei at time t,
N_0 number of nuclei at time $t = 0$,
λ_i decay constant for decay type I,
λ total decay constant.

It is not difficult to recognize that a partial activity does not depend on $e^{-\lambda_i t}$ but on $e^{-\lambda t}$. The reason for this is that the reduction in the total number of nuclei is the result of decays by all types of decays involved.

Now, what is the relationship between decay constant and half-life?—After one half-life period, half of all initial nuclei are left, after two half-life periods a quarter, after three an eighth and so forth. After ten half-life periods, the initial activity has been reduced to about 1/1000. To calculate the half-life, we have

$$1/2 = e^{-\lambda T_{1/2}} \tag{3.147}$$

and thus:

$$\ln(1/2) = -\lambda T_{1/2} \tag{3.148}$$

and

$$T_{1/2} = \ln 2/\lambda = 0.693/\lambda \tag{3.149}$$

Half-lives are specific constants of radio nuclides just as decay constants. Half-life is the preferred term in general usage. To present activity over time in a semi-logarithmic graph leads to a straight line (Fig. 3.36). Relative activity over time results in a curve as in Fig. 3.37.

Considering a single radioactive nucleus its lifetime may last between 0 und ∞. Its decay is not predictable. This is different for the determination of the average lifetime of many nuclei at the same time. In this case, we obtain a mean lifetime τ, obtained from the weighted sum of the life times of all nuclei with respect to their initial number. For the time dependency, one obtains

$$dN = \lambda N dt \tag{3.150}$$

and integrated:

$$N = N_0 e^{-\lambda t} \tag{3.151}$$

From this follows:

$$dN = \lambda N_0 e^{-\lambda t} dt \tag{3.152}$$

and for the total life time across all nuclei:

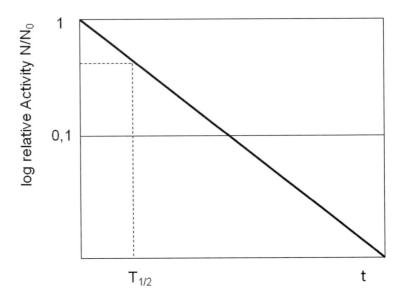

Fig. 3.36 Activity semi-logarithmic

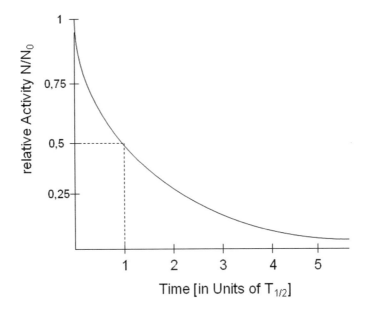

Fig. 3.37 Activity over units of half-life

$$L = \int_0^\infty tN_0\lambda e^{-\lambda_t}dt = N_0/\lambda \tag{3.153}$$

From this, the the mean life time is

$$\tau = L/N_0 \tag{3.154}$$

or

$$\tau = 1/\lambda \tag{3.155}$$

Half-life and mean lifetime thus have the following relationship:

$$\tau = T_{1/2}/0.693 = 1.44T_{1/2} \tag{3.156}$$

τ corresponds to that time, after which the activity has decreased to a value corresponding to 1/e (0.368) of the initial activity.

3.5.6 Reactions in Nuclear Physics

The most important nuclear reactions are
$_Z Y^N$ (n, γ)
(d, α)
(n, p)
(n, t)
(n, f)
(n, α)
(p, γ)
(d, p)
$(n, 2n)$
(α, n)

We will not discuss all those in detail here. We know nuclear fission and nuclear fusion. Besides them there exist a multitude of other reactions, which have been and are still employed to gain insights into the characteristics of atomic nuclei. In practice, one can shoot anything on everything.

As an example, I would like to return once again to fission since this plays a prominent role in social discourse. When $_{92}U^{235}$ fissions the following happens:

$$_{92}U^{235} + {}_0n_1 \rightarrow {}_{92-c}Y^{236\text{-d-m}} + {}_cZ^d + m_0n^1 \tag{3.157}$$

Fig. 3.38 Nuclear fission

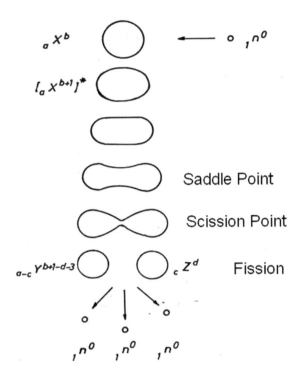

with Y and Z being fission product nuclei and m > 1. The latter fact is decisive to allow for a chain reaction within a sufficiently large (critical) mass of a Uranium isotope for example.

According to the liquid drop, model fission can be imagined as depicted in Fig. 3.38.

A chain reaction is necessary to liberate either continually or at a stroke a large amount of energy by nuclear fission, i.e. by a succession of nuclear reactions—in this case, fission—which is self-sustained until an available amount of fissionable material is used up (Fig. 3.39).

One of the pre-conditions for the functioning of a chain reaction is the availability of a sufficient critical mass. This means that so much fissionable material has to be available in a suitable geometry that the neutrons released do not pass the outer bounds of the body in question prematurely, which would disrupt the chain reaction.

3.5.6.1 Absorption

Besides fission cross sections absorption cross-sections play a prominent role in nuclear technology. They present a measure for absorption probability, i.e. the capture of a neutron, for example, $\sigma_{n,\gamma}$. There are materials with high absorption cross-sections just within the energy range, which is important for nuclear fission. These

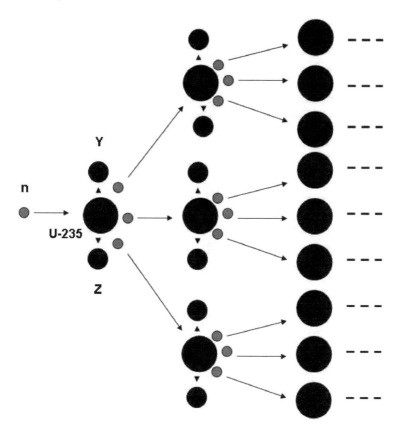

Fig. 3.39 Chain reaction

materials allow to control or stop chain reactions. Among them are cadmium and boron.

3.5.6.2 Moderation

It is important for controlled chain reactions to slow down the liberated neutrons to such energies, where the high fission cross-section of, for example, U-235 can become effective. This is the case in thermal equilibrium, at 0.025 eV. But since these neutrons originally have much higher energies they have to be brought to the necessary lower energy by scattering and thus successive energy transfer to nuclei of a suitable moderator. This is done most easily with light nuclei, for example, hydrogen. This is why water is used as a moderator in reactors. Carbon has also been used in the past.

Chapter 4
Types of Energy Generation

Abstract This chapter contains all major energy utilization technologies. They comprise steam power plants driven by coal, gas and oil and nuclear power, furthermore solar power plants and photovoltaic, wind power, bio mass and bio gas, geothermal heat, hydrodynamic power including barrages and pumped storage power plants, a proposal for rain power stations, combined heat and power generation and finally tidal power stations. For each energy carrier its origin is explained.

4.1 Steam Power Plants

4.1.1 Introduction

The transformation process for a thermal power plant functions in the following way: a fossil combustible releases its chemical binding energy during combustion. This energy can be used to generate mechanical work via suitable carrier media. In this section, we look at steam as an energy carrier. Steam in turn delivers a portion of its usable energy to a turbine driving a generator to produce an electric current.

Besides steam gas may be employed as a working substance as well (Sect. 4.1.3). In nuclear energy, nuclear binding energy replaces the liberation of chemical energy contained in fossil combustibles. We will learn more about the functioning of a nuclear power plant in a different section (Sect. 4.1.6).

In combustion engines, the combustion gas is used as the working medium by driving, for example, a piston motor before leaving it as exhaust fumes into the environment.

To characterize the energy transformation in a thermal or combustion plant, we usually make use of the energetic efficiency:

$$\eta = N/\overset{\circ}{m}_B h_u \tag{4.1}$$

with N the effective output of the plant, $\overset{\circ}{m}_B$ the mass flow of the respective fuel with the specific lower calorific value h_u.

These are some guide values of η:

- for older steam power plants 20–30%,
- for modern steam power plants up to 40%,
- for car motors about 25%,
- for large engines up to 42%,
- for open gas turbine plants 20–30%.

4.1.2   Coal

A coal power plant is shown schematically in Fig. 4.1. Other power plants with fossil energy carriers look fairly similar. Table 4.1 shows the exergy losses in a simple steam power plant.

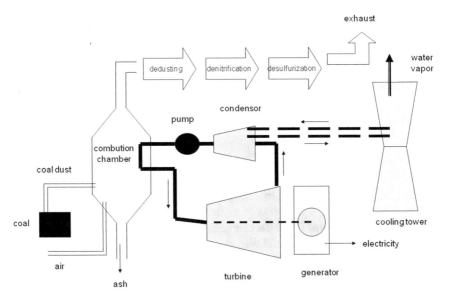

Fig. 4.1   Steam power plant

Table 4.1   Exergy losses in a simple steam power plant

Exergy loss by irreversibilities	Exergy loss relative to fuel exergy (%)
Combustion	31.0
Exhaust	4.0
Heat transfer	25.8
Turbine	6.9
Condenser	3.1
Feed pump	0.1

Concerning gas power plants, the following can be added:

There exist steam power plants fired by combustion gas. But there are also gas power plants employing a turbine driven by a different hot gas than water vapour. They are called gas turbine power plants (Sect. 4.1.3). And there are so-called gas-steam power plants. The designation gas power plant is also used for a combined heat and power plant fired by a combustion gas.

For the fossil energy carrier oil the same laws apply. However, oil is no longer employed for large-scale power plants. It actually continues to serve only as the emergency power supply for Diesel aggregates.

4.1.3   Gas

The gas in thermal power plants can be used in different ways. Classically the combustion of gas (fuel gas) offers itself as a substitute for other fossil fuels. From a physics point of view, the process is similar to the one described under Sect. 4.1.2. Technologically, however, it is of course different.

A further possibility is the employment of a gas turbine. In this case, a turbine is not driven by steam but by a hot mixture of air and combustion gas. This requires a construction, where the turbine has a combustion chamber as a front end (Fig. 4.2). By the relaxation of the gas mixture energy is transferred to the turbine blades converted into rotation energy. The efficiency of such a plant is about 40%. The rest is lost as heat transfers to the environment.

To increase the efficiency—for example, to 60%—a gas turbine can be combined with a steam turbine. Such gas-steam combination power plants use the waste heat of the gas turbine to drive the downstream steam turbine with the help of a steam generator (heat exchanger).

Combined heat and power generators are used for the simultaneous generation of electricity and heat. For this purpose, the waste heat, which is generated by the

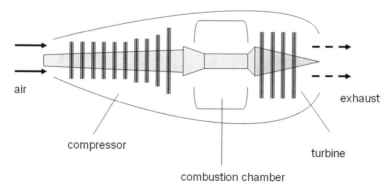

air

compressor

combustion chamber

exhaust

turbine

Fig. 4.2   Gas turbine

transformation to electrical energy, is used for heating. The mechanical power unit can be a combustion engine or a gas turbine. These combined units can be miniaturized to serve housing complexes or for example hospitals.

4.1.4   Oil

Today oil as an energy carrier does no longer play an important role. The principle is the same as for other fossil fuels. The energy released during combustion either drives a combustion engine or a thermal power plant. The remaining fields of employment for oil are emergency aggregates.

4.1.5   Thermal and Combustion Power Plants: Origin of Fuels

In thermal and combustion power plants fossil fuels are burnt. They can be coal, oil or gas. These fuels once came into existence by geological processes from organic matter. Organic matter in turn is the result of photosynthetic processes firstly enabled by the sun's radiation (Fig. 4.3).

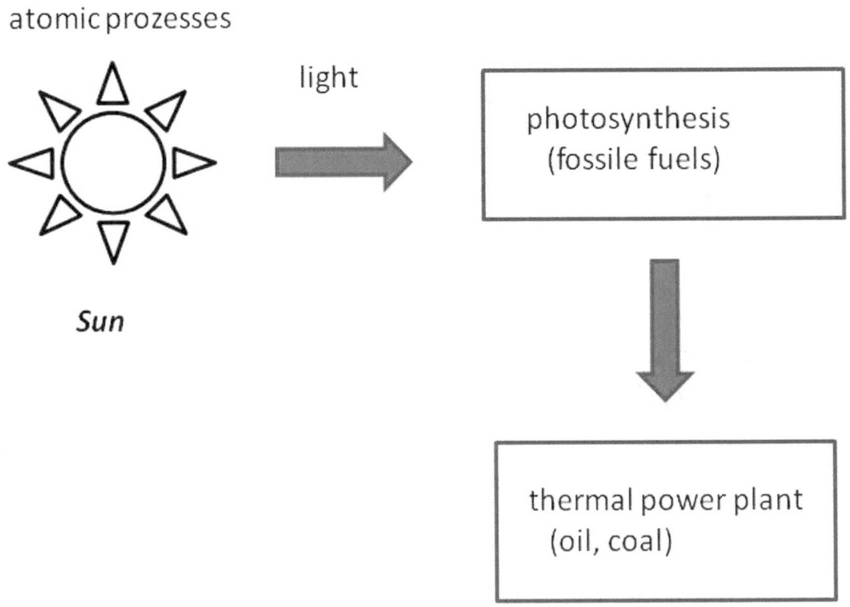

Fig. 4.3 Fossil energy

The combustion gases transmit their internal energy to a suitable medium, which is capable to drive a power unit, which in turn drives a generator transforming effective work into electric energy.

4.1.6   Nuclear Power

Different from fossil fuels, where chemical binding energy is set free, in nuclear energy the binding energy of the atomic nucleus itself is released by fission. Thus the combustion unit in steam power plants is replaced by a nuclear reactor, in which a chain reaction takes place. A nuclear reactor is supposed to generate controllable power by a stable operation.

This means a continuous succession of nuclear fission events per unit time. Fissionable materials contained in an isotope mix are used as fuel. They include U^{235} or Pu^{239}. The so-called multiplication factor k is used to control and design a reactor. It indicates the relationship between the neutron densities at the beginning and the end of one generation within a chain reaction. This factor has to be at least $= 1$ during the operation of a reactor. k is calculated the following way:

$$k = \varepsilon p f \eta L \qquad (4.2)$$

ε is the so-called fast fission factor, p the resonance escape probability; f indicates the percentage of moderated neutrons absorbed by the fissile material; η is the number of neutrons released during fission, and L is called the non-leak probability.

If we assume that during a single fission event the energy released is 180 meV, then 3×10^{10} nuclear fission events per second are required for power production of 1 W. The time-dependent power depends on

- the Volume V,
- the mean density N of fissionable nuclei,
- the fission cross-section,
- the mean neutron flux nv per cm^2 and second.

n and v indicate the mean density and velocity of the neutrons. Then the power of a reactor is

$$P\,[W] = nvN\sigma V/3 \times 10^{10} \qquad (4.3)$$

4.1.6.1   Characteristics of Reactors

Let us summarize briefly the main characteristics of nuclear reactors: The components common for most power station reactors are depicted in Fig. 4.4.

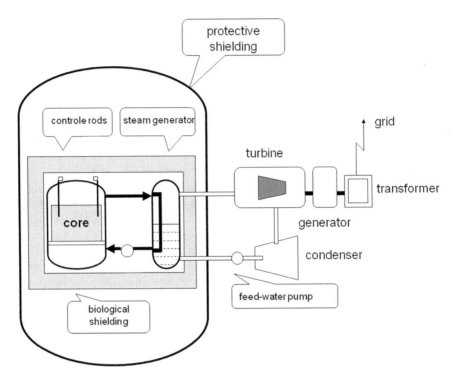

Fig. 4.4 Nuclear reactor

One of the most important characteristics of a reactor is the moderator. Heavy water, beryllium, graphite and water are the main moderators in use.

After the moderator, the coolant is the determining feature of power reactors. In high-temperature reactors, helium is used today as coolant and of course water serving at the same time as a moderator.

Today there are basically only two types of reactors: the gas-cooled graphite reactor and the water reactor, both deployed in large numbers. Besides them, there are liquid metal reactors as fast breeders. For water-cooled and moderated reactors, one distinguishes between pressurized and boiling water reactors.

4.1.6.2   Fourth-Generation Reactors

Presently, the discussion centres around the fourth-generation reactors, which are supposed to go into operation from 2030 onwards, among them molten salt reactors based on lithium fluorite as carrier salt. The latter have the advantage that no solid fuel pellets have to be produced beforehand and afterwards an online conditioning to separate fission products from the salt stream is possible.

It can be said generally that because of the high heat dissipation power by the necessary neutron flux density, reactors with liquid metal cooling are more suitable than those with classical coolants. Considered are, for example, liquid sodium, liquid lead or a lead–bismuth-eutectic. The ideal would be isobreeders, i.e. reactors, which breed as much new fissile material as they burn during operation.

The following fourth-generation types of reactors for fast neutrons are under consideration:

- sodium-cooled fast reactor (cooling by liquid sodium, uranium or MOX fuel elements; breeding factor: 1.2–1.45);
- lead-cooled fast reactor (cooling by liquid lead);
- gas-cooled fast reactor (cooling by helium; fuel elements from a Pu–U–Carbide mixture; breeding factor: 1);
- thorium molten salt reactor (mixture of thorium fluoride salt);
- fluoride cooled high-temperature reactor (fuel element beads in graphite blocs);
- super-critical light water reactor (cooling by water, fuel elements from oxide ceramics).

4.2 Solar Power Plants

As already indicated by their name: in solar power plants, electricity is generated from solar radiation by using a thermal intermediate step. In this case, one does not harness the light as such but the heat energy carried along solar radiation. But in any case in the end electricity is obtained from sunlight. Contrary to photovoltaic applications, which are employed as small decentralized units a solar power plant operates in a power range of several hundred Megawatts. Since thermal energy can also be buffered temporarily, the electricity generation of such power plants functions even during deficient or no solar radiation at all.

Solar power plants use a number of different technologies (Fig. 4.5).

The front end consists of concentrators, i.e. mirror configurations, which bundle the incoming sunlight and divert it. This is achieved by parabolic troughs or Fresnel lenses. The generated heat energy may reach temperatures of >100 °C. This heat energy is projected onto evaporation aggregates, generally tubes filled with a liquid of low boiling point or water. The steam, generated by using a heat exchanger, is then used in classical fashion to drive a turbine. Some plants use the water vapor directly to drive a turbine.

Solar power plants are employed preferably in the so-called sun belt of the earth, situated between 40° north and 40° south of the equator. There the probability of sunlight is high enough to justify their employment (the energy carrier here is directly available, such that transport and infrastructure costs, as they occur with other types of energy carriers, can be dispended with). Furthermore, the radiation intensity in this region is higher than elsewhere because of the angle of incidence of the solar radiation.

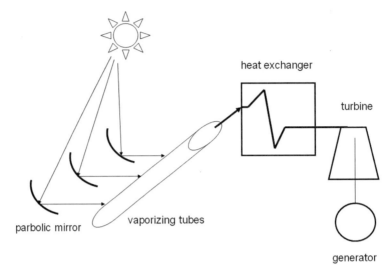

Fig. 4.5 Solar power plant

4.2.1 Parabolic Trough Power Plants

Parabolic through power plants, operated in particular in California, are well-tried. In Europe, one finds, for example, the Andasol power plant in the south of Spain consisting of several blocks. The total collector surface amounts to 1.5 million m². It is made up of several hundred thousand parabolic mirrors with a width of 6 m each and a length of 12 m. These parabolic mirrors bundle the sunlight and focus it on a receiver tube. This receiver tube is filled with special oil heated to about 400 °C. Steam generation is done via a heat exchanger. In addition, a thermal storage tank with a suitable liquid salt mixture (30,000 t) is available. This thermal capacity is able to operate the power plant for another 8 h, when the sun is not shining.

4.2.2 Photovoltaic

The sun radiates light and heat in the direction of the earth. Both parts are reduced from the original yearly 1.5×10^{18} kWh by the actual composition of the atmosphere (clouds, air moisture, dust, etc.). The light coming close to the surface of the earth or ships on the water can be harnessed to generate electricity with the help of the photo effect. To this end, solar cells are employed, normally embedded in solar modules. The totality of all interconnected solar modules is called a photovoltaic site.

The electricity generated in this way can directly be applied to any consuming terminal equipment or alternately fed into an existing electricity grid. When feeding in one has to take into account that solar cells initially generate direct current, but public transport grids function on the basis of alternating current. Therefore an inverter is necessary before feeding in.

Besides operating, for example, household appliances or feeding-in there are further possibilities for photovoltaic. It can be used to operate parking ticket machines or warning signs. To reduce their weather dependency, the electric energy is first used to load accumulators, which later operate the device.

The operation of photovoltaic sites is very environmentally friendly, since no combustion residues like ash or exhausts as in classical power plants remain. However, there are some disadvantages. During the production of solar panels chemical residues are created, which are problematic to dispose of. Furthermore, costs are significantly higher in comparison with conventional energy utilization technologies. In some countries, the widespread use of solar sites, for example, on the roofs of private homes has been made affordable because of subsidization by governments.

4.2.2.1 Efficiency

The efficiency is the quotient of electrical energy extracted within a given time span against the irradiated light energy. It can be calculated for different system aggregates:

- a naked solar cell,
- a solar panel or module,
- the overall configuration including inverter, charge controller and accumulators.

Efficiencies for solar cells in the context of photovoltaic attainable today are between a few percent up to 40%. Organic solar cells currently obtain up to 8.3% efficiency, thin-film modules based on amorphous silicon about 5–13%, thin-film modules based on cadmium telluride about 12%, solar cells from polycrystalline silicon 13–18%, cells from monocrystalline silicon between 14 and 24%. So-called concentrator cells can reach more than 40% under laboratory conditions.

A further problem is posed by weather conditions. The sun's radiation quite naturally not only fluctuates not only over the year but also during the day at the same location. Predictions are difficult. This means that because of its limited reliability reserve technologies are always necessary to compensate for power outages when the weather is unfavourable. In the long run, this problem can be solved with the help of suitable energy storage facilities. But until then solar energy remains just one component among others within an overall energy supply from different sources.

4.2.3 Solar Energy: Origin

Solar power plants use the heat from the sun's radiation, which is generated by nuclear fusion processes inside the sun.

For photovoltaic energy utilization, the atomic processes of the photo effect play a decisive role (Fig. 4.6).

Fig. 4.6 Solar energy

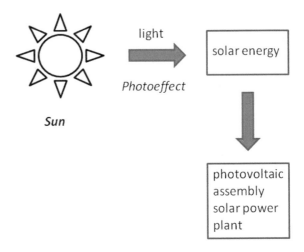

4.3 Wind Power

4.3.1 Introduction

Before we arrive at the proper applications, I would like to go one step back and start with the wind itself—its origin and the boundary conditions for its usage. These considerations will be followed by a supplement of physics foundations.

Thereafter we will come to the technical details. We will in detail deal with

- generation,
- rotor questions,
- losses,
- gains,
- feed in and control,
- offshore problems.

But now first of all some basic considerations:

Until now in our discussion, we have started out from quite different types of energy or energy carriers: binding energy (chemical and nuclear) as well as interactions in the atomic shell (photo effect). With the wind we encounter for the first time the necessity to transform kinetic energy directly into electricity (in power plants this happens in continuative steps via the kinetic energy of turbines). But now we have to deal with moving air masses.

Indirectly we have again to treat energy brought to us by the sun, because the movement of air originates from temperature differences in the atmosphere. Furthermore, we have to take into account that the warming of the earth's surface and thus also of the air masses does not happen uniformly but locally differently depending on a number of parameters:

- the incidence angle of the sun's radiation (temperature difference between polar caps and equator)
- atmospheric conditions (as already remarked in the preceding section about photovoltaic): transparency depending on air moisture, dust, etc.

The warming of the air does not occur directly but indirectly via the warming of structures of the earth's surface (mountains, land areas, formations created by men, etc.) by absorption of heat energy. Then the atmospheric layers above them grow warm by emissions from these structures.

All these processes are responsible for temperature and pressure gradients. Another important role is played by the change between day and night. This means that we have basically two dynamos being responsible for the massive movement of air masses and the generation of wind: the day-to-night rhythm and the differences between the poles and the equator. Other effects originate from the earth's rotation itself and the inclination of the earth's axis relative to the earth's orbit around the sun (summer/winter). The earth's rotation is responsible for the Coriolis force and thus for the formation of turbulences in the air movement. These eddies originating from airstreams from high pressure to low-pressure areas move anti-clockwise in the northern hemisphere and clockwise in the southern hemisphere.

Besides all this there are a number of local effects, which can lead to further temperature and pressure differences:

- differences between the warming and cooling of water and land,
- differences in the relief and the constitution of the earth's surface.

Problem 5: Calculation of Kinetic Wind Energy:

Let us take a random cross-section A, through which air passes during a time interval Δt with a velocity v, ΔV a volume element and ρ the specific weight. Derive the equation for the calculation of the kinetic energy of the air.

4.3.2   Realization

Now, which are the technical components, with which the physics of wind movement can be realized—or to put it differently: how does the energy transformation mechanism look like? For this, we have to look at the individual components of a wind power plant. It is composed of (Fig. 4.7):

- machine housing on a
- tower.

The rotor with its blades turns on a hub. The kinetic rotation energy is transferred to a generator inside the machine housing. Because the wind can change its direction

Fig. 4.7 Wind power plant

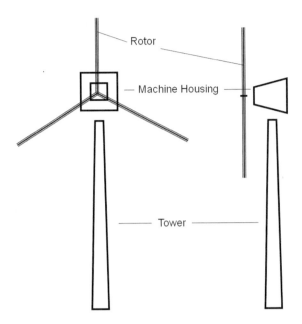

machine housing and thus rotor configuration are mounted revolvable on the tower. The necessary feed-in systems with their electronics are integrated partly in the machine housing or in a special switch component at the foot of the tower. Basically, this configuration enables the transformation of kinetic wind energy into electric energy, which can be fed into an electricity grid. In practice, however, there are some details to be observed:

A wind power plant initially generates alternating current. Because of the fluctuating rotational speed of the rotor in dependence on the wind velocity this alternating current in the state it is produced cannot be fed into a stable AC grid. Thus the current passes through a rectifier followed by an inverter, which takes care of the feed-in.

When constructing rotors, a number of criteria have to be taken into account, which will not be considered here in detail. They include

- optimal wind density,
- deceleration of the rotor in wind,
- vibration technology and stability (3-blade rotors are inherently more stable than 4-blade rotors, where different forces act upon the rotor blades depending on whether they are situated before the tower or above, in which case a lateral torsion moment would arise acting on opposing blades).

Altogether one tries to oppose a maximum of rotor surface against the wind.

The efficiency of the overall plant results from the efficiencies of all mechanical and electrical components combined.

Concerning wind energy, similar considerations apply as already stated in connection with photovoltaic: neglecting the fabrication of the individual components the energy transformation process is extremely friendly to the environment. Again there are no residues. However, here again, we depend on climatic conditions. The wind must not blow too weakly (<5 m/s) but also not too strongly, because otherwise, production will come to a standstill. This means as a consequence that a base load supply by wind energy is only guaranteed, if electrical surplus energy from storage media is held in reserve for bad conditions, meaning that wind power plants as well can only be part of an optimized energy mix.

Wind power plants are placed in such locations, where optimal wind conditions can be found—basically on hills and at the coast or on flat land, where no close building constructions disturb it. The electricity gained depends of course on the location itself. Plants at the coast deliver about 25% more energy than plants inland. Offshore plants deliver about the double amount against inland plants. Concerning offshore plants specific cost and maintenance aspects have to be taken into account (in addition to sea cables to the coast). Because of the corrosiveness of sea water and air layers above it, there is a danger of corrosion, so that special construction materials have to be used. In addition for certain components, special protective measures are required. And here are some further conditions to be observed for offshore plants:

- higher wind velocities,
- vibrations of the whole plant due to the motion of the sea,
- maintenance accesses (landing places, helicopter platforms).

4.3.3 Wind Energy: Origin

To generate electrical current by wind power plants the kinetic energy of moving air masses is used. The movement of these air masses originates from the adjustment of differently warmed atmospheric regions. Temperature differences originate from uneven sun radiation onto earth as already described above. At the end, in this case also nuclear fusion and the resulting heat inside the sun are the original causes (Fig. 4.8).

Fig. 4.8 Wind power

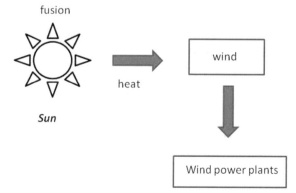

4.4 Bio Mass

Although there are significant differences concerning the origin of the energy carriers as well as their transformation into usable energy between them, bio mass facilities do not so much differ from conventional power plants using fossil energy carriers: organic material is combusted, and the chemical binding energy released during the combustion process is used to drive mechanical-electric components.

Let's first have a look at the energy carrier. What is called bio mass, is anything that grows and decays somehow. Of course for our selection, certain criteria have to be met excluding a number of materials. Important aspects are calorific value, storability, transport, availability, costs, etc. This leaves us with

- vegetables,
- fruits
- garden waste,
- leftovers,
- but also straw and fruit pods.

For major bio mass plants these are more suitable:

- dry wood,
- wood wast,
- chips, etc.

Normally the fuel has to be processed, i.e. brought into a shape, which can be stored and transported and is suitable for a major combustion plant. Quite often bio mass is combusted together with fossil energy carriers. Since, as already mentioned, the combustion processes are similar, the possibility exists to convert existing power plants accordingly.

4.4.1 Combustion Technology

The combustion technologies are different for small units for domestic utilization with respect to major plants. Sometimes it is difficult to draw the line. Basically, all furnaces, in which wood or wood waste are burnt, can be called bio mass sites. Today these predecessors have been refined technologically and upgraded as a pellet or automatic woodchip heaters.

Bio mass plants are classified according to the following criteria:

- total size of the plant,
- type of fuel,
- loading system,
- combustion technology.

Table 4.2 Classification of bio mass plants along with plant size

Size	Power
Very small plants	<15 kW
Small plants	15 kW–1 MW
Medium size plants	1 MW–50 MW
Large-scale plants	>50 MW

Concerning plant size, Table 4.2 gives some indications:
Loading systems can be differentiated with respect to

- continuous,
- step wise,
- manual,
- automatic loading.

Combustion technology is differentiated according to

- underfeed stoker,
- grate,
- inclined grate,
- pusher grate,
- fluidized bed combustion,
- dual-chamber furnace,
- direct firing system,
- shaft furnace,
- top burning furnace,
- bottom burning furnace,
- front fire furnace.

We will not go into the details of these differences here.

The main components (Fig. 4.9) of a bio mass combustion plant (as a front end before the transformation into usable energy) are:

- fuel storage,
- fuel loading,
- transport unit,
- boiler,
- air control,
- fuel gas cleaning,
- deashing,
- heat exchanger.

Figure 4.10 shows a section of the fuel transport cycle.

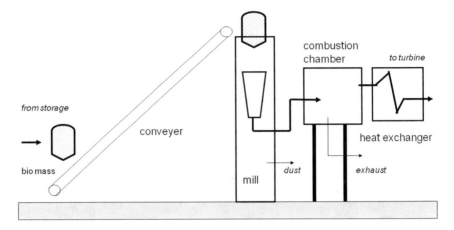

Fig. 4.9 Bio mass combustion plant

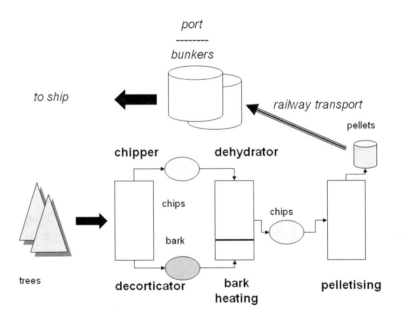

Fig. 4.10 Fuel transport cycle of bio mass

4.4.2   The Combustion Process

During combustion, one distinguishes various phases:

- dehydration (release of water vapour),
- pyrolysis (release of combustible gases),
- oxydation (gas combustion).

All three phases require different oxygenations.

During the whole process, the following substances are released or created:

- water vapour,
- carbon dioxide,
- carbon monoxide,
- nitrogen oxides,
- hydrocarbons,
- smut,
- cinder.

Factors influencing the combustion and thus also the efficiency of such plants are

- calorific value,
- fuel moistness,
- oxygenation,
- gas performance in the combustion chamber,
- combustion temperature,
- combustion chamber layout.

The process to release biochemical binding energy executes in several steps. The rate of oxygenation is used to distinguish the different phases and procedures concerning thermal usage.

In the overall process, bio mass is transformed to water vapour, carbon dioxide, carbon monoxide (small amounts), hydrocarbons (small amounts) and smut (small amounts). Residues are cinder and ashes.

The combustion process passes through three phases. The first phase is dehydration, in which nearly only water vapor is released. It follows pyrolysis, where combustible gases like hydrocarbons, carbon monoxide and hydrogen are set free. Finally, there is oxidation by burning the gases created during pyrolysis.

A complete combustion is achieved by high exploitation of the fuel energy and low emissions. In contrast, by imperfect combustion still combustible gases and dusty fuel particles are released.

Deficiencies during the thermal utilization of bio mass result from radiation (depending on the construction 3–7%) and exhaust (depending on fuel, the fraction of carbon dioxide in the exhaust and the temperature ratio between exhaust and air). Efficiencies vary between 60% (log wood boiler) and 90% (automatic wood chip furnace) relative to the heat output. In contrast, the efficiency of an open fire is about 5%.

4.4.3   Fuel

When selecting (new) fuels, the following criteria are important:

- chemical composition,

- danger of explosion,
- ease of processing,
- combustion process.

Under these criteria, the following materials have been accepted into the bio mass cycle:

- wood pellets,
- palm pulp pellets,
- citrus pellets,
- peanut pellets,
- rice husks,
- soybean husks,
- coffee husks.

Because of their different densities, wood chips have a somewhat lower, wood pellets a somewhat higher calorific value than traditional coal. 60% of wood pellets come from Canada (by boat), 10% from Europe and the remaining 30% from other locations (source: RWE) with the result that meanwhile, the transport routes resemble those of large oil tankers.

4.4.4 The Generation of Electricity and Heat

Until now we have dealt only with front-end considerations. But in the end, utilizable output is the final aim. Large energy utilities operate many combined heat and power plants driven by bio mass all over Europe. They function in the following way:

- Bio mass is delivered to the power station and foreign substances like stones and metal are sorted out.
- Thereafter the fuel is fed into a boiler and is combusted at high temperatures (as explained above).
- The energy released heats water to evaporate. The vapor drives a turbine which is coupled to a generator. In this way, electricity is generated.
- The waste heat from the boiler can be used for district heating purposes.

4.4.5 Bio Mass: Origin

There are two ways for energy utilization of biological substances. In this section, we have discussed the utilization of bio mass. Similar to fossil fuels bio mass is the end product of photo synthesis enabled by the sun's radiation (Fig. 4.11).

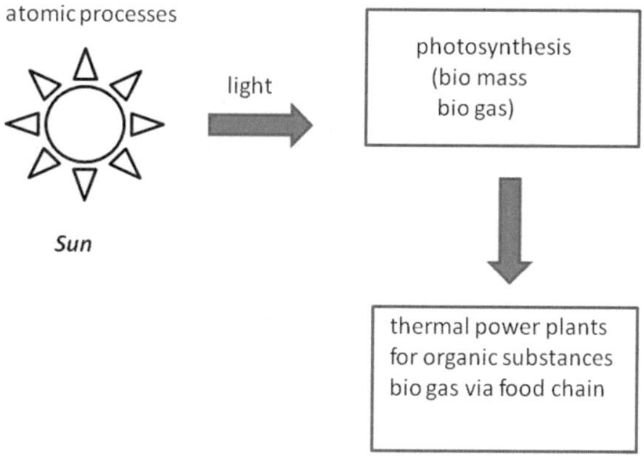

Fig. 4.11 Bio mass, bio gas

4.5 Bio Gas

The energy carrier for bio gas in energy utilization is methane. During bio gas generation some other gases are present, but they are only relevant concerning efficiency. Bio gas is harvested as a by-product in farms, from wastes of the food industry and wastewater treatment plants. The production rate depends of course on the type of substances employed. These can have the following origins:

- manure and slurry,
- weeds and herbal waste,
- residues from distillation and brewery processes,
- organic mire from food production,
- slaughterhouse wastes,
- sewage sludge from municipal plants,
- food wastes from restaurants,
- other garden wastes,
- special cultivated plants with high yield.

Table 4.3 summarizes gas yields for different organic substances.

The basic process responsible for everything else is the decomposition of organic material be it dead plants or animal carcasses. This process is enabled by bacteria with no need for oxygen to live on (anaerobic). Such bacteria are prevalent in nature, for example, in swampy areas, and they are located in the digestive tract of animals. From farm cattle, they end up in manure and slurry. This is the chemical reaction:

$$\text{organic waste} + \text{bacteria} \rightarrow \text{methane} + CO_2 + \text{decomposition product}$$

Table 4.3 Bio gas yields
(*Source* T. Seilnacht)

Substance	Yield (m^3/t)
Cattle slurry	25
Pig slurry	36
Whey	55
Brewery residues	75
Garden garbage	110
Bio waste	120
Food residues	220
Used grease	600

4.5.1   Composition and Characteristics of Bio Gas

Bio gas is a mixture consisting of methane (CH_4) (50–70%), carbon dioxide (CO_2) (30–40%) and traces of hydrogen sulfide (H_2S), nitrogen (N_2), hydrogen (H_2) and carbon monoxide (CO) (Table 4.4). The average calorific value of bio gas amounts to 25,000 kJ/m^3. Thus the average calorific value of one cubic metre bio gas corresponds to about 0.6 l fuel oil. Table 4.5 lists some characteristics of bio gas:

Problem 6: Calculate the annual equivalent of fuel oil produced by a cow.

Table 4.4 Bio gas
composition (*Source* T.
Seilnacht)

Gas	Fraction (%)
Methane	40–75
Carbon dioxide	25–55
Water vapour	0–10
Nitrogen	0–5
Oxygen	0–2
Hydrogen	0–1
Ammonia	0–1
Hydrogen sulphide	0–1

Table 4.5 Some
characteristics of bio gas

Parameter	
Density	1.2 kg/m^3
Calorific value	4–7.5 kWh/m^3 (depending on methane content)
Ignition temperature	700 °C
Ignition concentration gas content	6–12%
Smell	Because of hydrogen sulphide: rotten eggs

Manure production of one cow per day: 10–20 kg.
Conversion into bio gas: 1–2 cm³/d.
Conversion into electricity: 1.2–2.4 kWh.
With the average calorific value given above, calculate the annual equivalent of
fuel oil produced by one cow.

4.5.2 Technological Pre-conditions

These are the major components of a bio gas plant (Fig. 4.12):

- fermenter,
- combustion unit,
- electricity generator.

In addition, a post fermenter can be placed between the fermenter and the final storage container (not displayed), in fact, 1–2 per fermenter.
The process to generate electricity or heat runs as follows:

- input of the organic substances into the fermenter,
- dwell duration: several days → creation of bio gas,
- during dwell: continuous stirring,

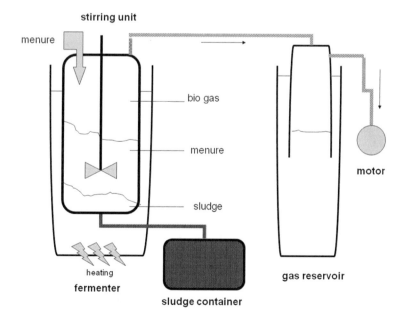

Fig. 4.12 Bio gas plant

- dwell duration and volume generate bio gases depending on the temperature (30–40 °C attempted),
- transfer of the remaining mass into the final storage container,
- there again recovery of rest bio gas,
- gas cleaning by oxygen; desulphurization,
- dehydration of the gas in a separator.

Although Fig. 4.12 distinguishes functionally between fermentation tank and gas storage, the gas storage may be placed above the fermentation tank as an extensible membrane, from where the gas is led directly to the motor to be operated.

After this the combustion can start in different aggregates

- combined heat and power plant,
- engines zu generate electricity,
- usage of waste heat during the combustion process (heating of the fermentation tank),
- feed into the public power supply.

The energetic origin of bio gas is similar to the one for bio mass already explained in Sect. 4.4.5.

4.6 Geothermal Heat

4.6.1 Introduction

This section treats an energy carrier, which is again supposed to have indefinite potential: geothermal heat. We will differentiate between simple systems for domestic use and larger power plants.

Let us first come back to the subject of the heat pump. This has to do with the ground soil and its resources and what types of system solutions are generally available.

Further down, we will explain the basics of geothermal energy and the functioning of geothermal power stations.

As already mentioned there are two main areas of application:

- small units assisted by simple heat pumps,
- actual power plant operations.

Both applications differ considerably from each other with the exception of the energy carrier itself. Independent from this the main attention is on the enormous differences in depth that have to be reached to obtain the desired results. For small applications, we talk about near-surface systems, while deep drilling has to be employed to unlock the real geothermal potential.

4.6.2 Heat Pump Systems

The heat pump, its energy balance and functioning have been discussed in Sect. 2.2.6.1.

For their employment, only layers of earth between a depth of 1.2 and 100 m need to be dug up. What do the technical implementations look like? There are two possibilities:

- deployment of ground collectors in depths of up to 1.2 m,
- deployment of geothermal probes, for which boreholes down to 100 m of depth are required.

Both technologies require carriers capable of absorbing geothermal heat. This can be a liquid, for example, a mixture of water and glycol. The heat power generated depends of course on the geological environment as well as on the overall geometry of the plant itself. Ground collectors offer themselves for new family homes in particular, because collectors can be implemented already during the building phase.

For larger buildings and in case, the earth conditions do not favour collectors, geothermal probes are the next choice. Boreholes go down as deep as 100 m with a diameter of 20 cm. Generally, several probes have to be deployed. In 100 m depth, the average temperature remains constant at about 12 °C independent of the season. However, the overall energy balance requires that a heat pump needs an electrical current from an outside source in the first place to power it.

4.6.3 Geothermal Power Plants

Geothermal Power Plants utilize so-called deep thermal energy. To unlock these energy resources boreholes down to 2000 m and more are required. Because of the high costs associated with it, this technique pays off only for large projects. Energy carrier is hot brine, which is pumped upwards. The brine reaches temperatures of nearly 100 °C. This heat can be used for district heating or transformed into electric energy. In the latter case, special turbines are required driven by a medium of low boiling point.

Geothermal power plants cannot be erected at any arbitrary location. A suitable geothermal reservoir has to be found first of all. Furthermore, the rock bed has to allow for economic drilling.

4.6.4 Geothermic Energy: Origin

It is well known that the earth itself is warmer than it should be from the sun's radiation alone (this does not refer to the atmosphere but to the temperature of the

radioaktive decay

Fig. 4.13 Geothermal energy

ground soil itself). The cause is radioactive decay of unstable nuclei in the interior of the earth. Here the laws of nuclear physics apply. Radioactive decay releases energy causing the interior of the earth to warm up (Fig. 4.13).

4.7 Hydrodynamic Power

4.7.1 Introduction

There are two methods to utilize water for the generation of electricity:

- placement of turbines in barrages in running waters,
- as pumped storage power plants.

Pumped storage power plants are the subject to recent discussions, because they are supposed to be located in mountainous regions to transform electricity generated by wind power at the coasts into potential energy there. The aim is to transport electricity from them to major cities. However, one has to consider conduction losses over long distances.

4.7.2 Barrages

At the entry of the plant, a grate holds off floating refuse. Pieces of wood, twigs and other waste are automatically extracted from the water and transported to be sorted, composted and for other purposes. The water reaches the turbine housing, drives the rotor and puts the transmission shaft into motion. The shaft drives a generator to produce electricity. A turbine type with as high an efficiency as 80–95% is the Kaplan tubular turbine. These turbines are installed with a small inclination against the horizontal in the direction of the water flow. The water flows around a watertight bulb-like steel casing, containing the rotor, gearing and the generator. At the end of the casing, the water encounters a ring of 20 guide blades. They direct the water in such a way that it drives the rotor with its four blades at a favourable angle. Thus the rotor of a Kaplan turbine resembles the propeller of a ship. Because the guide blades and the main blades can be adjusted the turbine can be tuned to seasonal variations of the water level. Water volume and drop height are the deciding factors for turbine power. Electricity can only be generated for a drop height of 2 m or more.

4.7.3 Pumped Storage Power Plants

Pumped Storage Power Plants utilize the potential energy of water to generate electricity. When transforming potential into kinetic energy the drop height is decisive. In mountainous regions, barrage dams can be constructed to retain water from the flow of one or more streams. In this way in the high mountains plants can be constructed consisting of several barrages with their basins connected via ducts to obtain total drop heights of several hundred meters and up to 1 km.

In this way, major power plants can be built generating several thousand MW of power with the appropriate turbines and generators.

These pumped storage power plants are really energy buffers to be deployed to smooth energy demand. First of all, electricity originating from other sources (for example, wind power plants in a flat country) is used to drive pumps feeding water from a lower basin into a higher one (Fig. 4.14). In this way, the water in the upper basin gains potential energy. In times of high demand, this potential energy can then, in turn, be converted into kinetic energy by driving generators after releasing the water from an upper basin to flow down into a lower one.

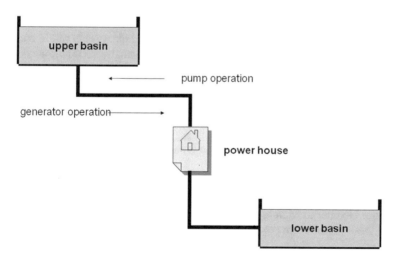

Fig. 4.14 Pumped storage power plant

4.7.4 Hydrodynamic Power: Origin

To come back to the origin of the kinetic energy carried by flowing water (kinetic energy from gravitation downhill), one has to look at the water cycle of the earth. To put it simply, the sun heats the earth, water evaporates, gathers in clouds, which bring back the water to earth as rain under certain conditions and fills up our rivers. In detail, this process is much more complex. We shall not discuss factors like aerosols, temperature gradients in the atmosphere, etc., responsible for the formation of clouds as well as sufficiently heavy water droplets within a cloud, which then fall to earth, here.

4.8 Rain Power Stations

4.8.1 Background

Hydrodynamic power plants utilize the potential energy of barrages or pumped storage power plants that retain water masses. The water masses are normally taken from streams or water reservoirs replenished in turn by rainfall. Theoretically one could do without the interposition of such plants, if one utilizes the kinetic energy released during rainfall directly. In this way, one could save investments in barrages and in retaining technology on the one hand, and on the other, the energy transformation could happen in a direct way and thus with a smaller increase in entropy.

Such a possibility would be realized by a rain power plant. This would constitute a completely new type of energy utilization, which actually (2021) does not exist in the world at all. The proposal is put forward for the first time in this book. Its usability has yet to be proven in practice.

Rain power plants do only make sense, when heavy rainfall continues for a long time span, which means that they would not be economic in temperate zones like Europe. However, they could satisfy local energy demand in rural areas for a sustained period of time. This would be the case in countries belonging to the so-called rain belt of the earth.

The tropical rain belt oscillates between the northern and the southern tropic. The rainy season in the southern hemisphere of the Indian Ocean and the West Pacific lasts from October to March, in the northern hemisphere from April to September. During these periods rain power plants can be operated economically.

4.8.2 Concept

Figure 4.15 shows a schematic presentation of a rain power plant.

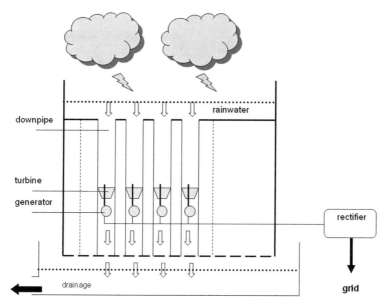

Fig. 4.15 Rain power plant

The power plant consists of an array of downpipes with a small turbine placed in the lower third in each pipe driving a generator or dynamo. A basin above the pipes is filled with rain water. Once the water level reaches a certain mark and critical pressure builds up the pipes are opened, water runs down, drives the turbines, which generate electricity with the dynamos, which is collected and fed into several rectifiers, then fed into the local grid with the appropriate alternating current frequency via inverters. The water then is discharged below the power plant block.

4.9 Combined Heat and Power Generation

A further possibility to utilize energy efficiently is combined heat and power generation. This technology constitutes a combination of generating electricity together with heat at the same time. In such plants, several of the technologies already discussed can be employed:

- steam power plants,
- gas turbines,
- combustion motors, etc.

Different substances as energy carriers are possible as well:

- gas,
- coal,

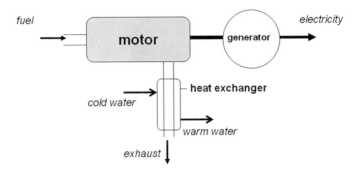

Fig. 4.16 Block-type thermal power plant

- oil,
- bio mass.

Well-known are block-type thermal power stations. If electricity generation is driven by a combustion engine, the waste heat produced is used for heating in the cold season (Fig. 4.16).

4.10   Tidal Power Stations

The mouth of the river Rance near Saint-Malo on the northern shores of Brittany in France offers a magnificent sight—but not only because of the landscape, but also because of a technical construction that searches its equivalent elsewhere in the world. If one takes a trip by pleasure boat into the entrance of the water gate, initially nothing indicates the presence of this technically large-scale project. Once the movable bridge above the water gate is raised traffic comes to a standstill on the mighty levee, which connects Saint-Malo with Dinard in northern Brittany.

The secret of the technical machinery in question is not visible at first sight but hidden below the surface: the Tidal Power Station at the Rance. But even the impressive view into the hall inside the dam below the roadway does not necessarily disclose the purpose of the installation. However, there exists a multimedia exposition area of 300 m² about the tidal power station, which initiates the visitor free of charge into the secrets of the energy transformation by tidal forces: the first tidal power station worldwide, constructed between 1962 and 1965. The major project on the northern shores of Brittany delivers 600 Million kWh per annum in the form of electrical energy and can thus supply alone about 300 000 inhabitants with electricity.

The electricity is created by 24 mighty generators, each of which can produce a maximum power of 10 Million Watt. They are driven by water currents induced by the tides two times a day in one direction (from the sea into the mouth of the Rance) and two times a day in the opposite direction (back to the sea), passing through the turbines: day after day.

The location of the tidal power station had been chosen on purpose. The tidal difference in the mouth of the Rance belongs to the biggest on earth and can reach 14 m. The reason for this is the low depth and the backwater effect of the channel.

Meanwhile, other countries have followed the Rance example. Today, there exist tidal power stations in Canada, China, Russia, South Korea and the UK.

Chapter 5
Storage Technologies

Abstract The distribution of electricity is a decisive factor in power industry. It is achieved by the operators of transmission grids and distribution networks. Ideally electricity should be generated, where it is required. This is not always possible, because of climate conditions. Thus one arrives at a power mix with two types of challenges:

- intermediate storage for bottleneck situations,
- transport to non-producing areas.

Intensive research is done to find storage technologies to circumvent the problem of conduction losses by positioning storage facilities near generating plants. Electrochemical storage is used in the mobile sector (portable devices needing an energy source, electric vehicles) as well as in the stationary sector (emergency generators)..

The distribution of electricity is a decisive factor in the power industry. It is achieved by the operators of transmission grids and distribution networks. Ideally, electricity should be generated, where it is required. This is not always possible, because of climate conditions for example. Wind power plants are more efficient near the coast or on hills than in calm valleys. Furthermore, massive sun power is generated during advantageous irradiation, while demand in the area could rise at night. Thus one arrives at a power mix with two types of challenges:

- intermediate storage for bottleneck situations,
- transport to non-producing areas.

Both challenges are interconnected.

5.1   Transmission Losses

5.1.1   Voltage Levels

The voltage with which electricity is generated in large power plants is between 10,000 and 21,000 V. In this respect, it is therefore not suitable for the end user, on the other hand, this voltage does not suffice to equalize the incurred conduction losses. Therefore, the voltage is transformed to a much higher level. These voltage levels are different in different countries. Altogether four different voltage levels are in use:

- extra-high voltage (220–1150 kV),
- high voltage (110 kV),
- medium voltage (bis zu 20 kV),
- low voltage (100–230 V).

The extra-high voltage grid is operated nationally and internationally. The high-voltage grid is used to serve large industries and areas of high population density. The medium voltage level finally feeds transformers responsible for distribution to households and small businesses at low voltages levels.

5.1.2   Conduction Losses

Conduction losses originate along lines during transmission within electricity grids. They depend on

- length,
- cross section,
- conductor current,
- outside temperature,
- initial voltage,
- conductor resistance.

Conduction losses amount to about 6% per 100 km for a 110 kV line and could be reduced to about 0.5% per 100 km at 800 kV extra-high voltage lines.

5.2   Storage

Intensive research is done to find storage technologies to circumvent the problem of conduction losses by positioning storage facilities near generating plants. One type of storage power plant has already been discussed in Sect. 4.7.3.

Electro-chemical storage is used in the mobile sector (portable devices needing an energy source, electric vehicles) as well as in the stationary sector (emergency generators). Their employment in connection with transmission grids and distribution networks is not envisaged. The following types of batteries are in use:

- lead-acid batteries,
- lithium-ion batteries,
- lithium-sulphur batteries,
- lithium-air batteries,
- sodium high temperature batteries.
- Redox-flow batteries.

5.2.1 Lead-Acid Batteries

This type of battery uses as electrolyte diluted sulphuric acid in a casing together with a pair of electrodes made of lead and lead dioxide. As a single unit, they possess a nominal voltage of 2 V. They have been in use for more than 100 years. They are employed in uninterruptible power supplies and may serve to support small isolated stationary networks. Mobile, they are used as starter batteries or to drive floor borne vehicles.

5.2.2 Lithium-Ion Batteries

This type of battery is based on the exchange of lithium ions between anode and cathode. The cathode material consists of inorganic transition metal oxides or mixed oxides. Graphite is used for the anode. The electrolyte is lithium salt dissolved in an organic liquid. Nominal voltages are a few Volts. Their main fields of employment are the operation of portable communication devices and electro-mobility because of their high energy density.

5.2.3 Lithium-Sulphur Batteries

This type of battery promises high energy density as well. The anode is made of lithium and the cathode of sulphur and carbon. When discharging, the lithium oxidizes. The lithium ions migrate into the electrolyte. They then combine with sulphur to lithium sulphide at the cathode.

There are still obstacles against commercialization (2021) because of security aspects concerning metallic lithium and its contact with moist air and regarding its cycle stability. The advantages of higher energy density are compensated though by its relatively high volume at the same time making deployment in electric vehicles problematic for example.

5.2.4 Lithium-Air Batteries

This type of battery uses oxygen as an oxidation medium. For this purpose, air is added to the cathode. The lithium ions of the anode combine with lithium peroxide. The energy density is small but also the volume required. They are still in the experimental phase. Security aspects are the same as for lithium-sulfur batteries.

5.2.5 Sodium High-Temperature Batteries

These batteries are operated at temperatures between 290 and 350 °C. At these temperatures sodium is liquid. The electrodes are liquid sodium and liquid sulphur, while the electrolyte is a ceramic solid body (sodic aluminum oxide). The liquid state of the electrodes is kept because of the charging and discharging processes. Longish downtimes, however, are not possible because of occurring solidification.

Liquid sodium, however, harbors high-security risks, since it would react immediately when in contact with air. Therefore, these batteries are not suited for operating electric vehicles. But they are employed in networks for purposes of load and supply management.

5.2.6 Redox-Flow Batteries

This type of battery uses the electrolyte as an energy store. A reaction unit is placed in its centre divided by an ion conduction membrane into two halves. Each half contains an electrode. The electrolyte is contained in two external reservoirs (salts dissolved in acid). The charging levels of the electrolytes are different in both reservoirs. For discharging, the electrolytes are fed into the different halves of the reaction unit. This enables an ion exchange reaction via the membrane. The electrolytes are based on vanadium, vanadium bromide, sodium bromide, zinc bromide or iron-chromium.

These are the advantages of this type of storage:

- power and capacity can be fixed independently from each other,
- long working life,
- no fire risk,
- simple storage of high capacities using only a few subsystems,
- low self-discharging,
- easy recycling.

Disadvantages:

- uneconomical for applications demanding a great deal of power and little energy,
- uneconomical for small applications,
- unsuitable for electric cars,
- lower efficiency in comparison to lithium-ion batteries.

Chapter 6
Future Approaches

Abstract Besides those technologies discussed up to here, there are further approaches in the state of development that have as yet not been proven in practice on a significant scale to solve energy utilization problems. Two of those shall be briefly described:

- nuclear fusion,
- fuel cells.

Besides those technologies discussed up to here, there are further approaches in the state of development that have as yet not been proven in practice on a significant scale to solve energy utilization problems. Two of those shall be briefly described:

- nuclear fusion,
- fuel cells.

6.1 Nuclear Fusion

Every so often reports about the utilization of nuclear fusion are being produced, so recently about the ITER project in France. ITER has one thing in common with the ISS and CERN: these three projects drive the most expensive machines ever thought out by humankind.

Ever since the first hydrogen bomb had been ignited we know that all theoretical predictions had been fulfilled: in nuclear fusion of hydrogen isotopes binding energies are released, which can be utilized one way or another:

$$H^2 + H^3 \rightarrow He^4 + {}_0n^1 + 17\,\text{MeV} \tag{6.1}$$

$$H^2 + H^2 \rightarrow He^3 + {}_0n^1 + 3\,\text{MeV} \tag{6.2}$$

$$H^2 + He^3 \rightarrow He^4 + {}_1p^1 + 18\,\text{MeV} \tag{6.3}$$

The energy released during fusion is higher by orders of magnitude than that during nuclear fission. But the technical challenges for controlled continuous fusion are so enormous that the time span, which could be surpassed in experimental machines was only extremely short (<0.5 s). A reactor in the true sense of the word does not exist today, not to mention a basis for commercial utilization.

6.1.1 Challenges

The objective is to divert the energy released during controlled nuclear fusion in such a way that electricity generators can be driven with the help of heat exchangers. This approach, followed since the early 50 s of the twentieth century, promises the unlocking of immeasurable energy resources.

Those would be chemically clean with the production of little radioactivity with significantly shorter half-lives in comparison to conventional nuclear power plants. However, as of today, we are still a long way short of reaching this goal. The main problem is still the control of nuclear fusion itself.

There are two directions of research dealing with practical fusion machines:

- the TOKAMAK principle,
- stellarator technologies.

ITER (International Tokamak Experimental Reactor) follows the first path. The challenges for this method are:

- inclusion of a plasma of 100 million degrees for a sufficiently long time span to enable continuous fusion
- development of suitable structural materials resisting the emerging neutron flux long enough so that these components do not have to be exchanged after only a short period of operation
- production of net energy output higher than the one applied to make the reactor work in the first place.

Until now, one has not succeeded to maintain a stable fusion plasma for more than half a second. And net output still remained far from the breakeven point.

ITER is a collective project with the participation of the EU, the USA, Russia and other countries. The device is being built in Cadarache in France and will deliver the first results in about 10 years' time. Until then more than the estimated sum of about 10 billion Euros will be used up. ITER is supposed to solve the problems mentioned. Its construction was finished by 50% in 2017. The first plasma with a pulse duration of 400 s is supposed to be generated in 2025.

But despite that ITER would still not be a reactor. This shall be demonstrated by the follow-up machine DEMO—as the acronym indicates. Thus fusion research means a lot of patience and the willingness of the international community to continue to invest a lot of money in the future. This factor can be decisive concerning the future of fusion technology besides the technical problems themselves.

6.1.2 Technical Questions

A Tokamak fusion reactor poses other specific challenges to the engineers—apart from the fusion physics itself. One of the problems, as already mentioned, concerns material resistance. The neutrons released during the fusion process affect the walls of the reactor. This led to the search and development of suitable materials to keep the exchange of brittle and porous structures at a minimum. Furthermore, radioactive by-products as well as bred tritium need appropriate handling and disposal.

Concerning fuel availability, deuterium is present in hydrogen compounds and in the seas in large quantities and can be obtained relatively easily. With tritium it is different. Tritium decays as a β-emitter with a half-life of 12.4 years and is present in natural hydrogen only in very small quantities. However, it can be produced from lithium using the following reactions:

$$Li^6 + {}_0n^1 \rightarrow He^4 + T + 4.78 \text{ MeV} \tag{6.4}$$

$$Li^7 + {}_0n^1 \rightarrow He^4 + T + {}_0n^1 - 2.47 \text{ MeV} \tag{6.5}$$

Figure 6.1 shows a future fusion reactor. Its central element is the plasma chamber. The fuel (deuterium, tritium) is fed into the plasma chamber for fusion. Further plasma heating is done by He ions. The neutrons generated during the fusion reactions hit the reactor walls. This leads to other nuclear reactions. The binding energy released during fusion heats a coolant, which renders its heat in a heat exchanger to generate steam. Thereafter energy transformation takes place in quite classical fashion by driving a turbine and a coupled generator. Only helium remains as

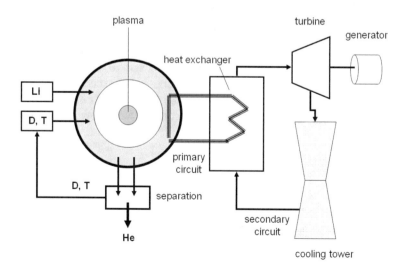

Fig. 6.1 Fusion reactor

Table 6.1 Technical parameters Wendelstein X-7 (*Source* Max-Planck Institute for Plasma Physics, Greifswald)

Parameter	
Large plasma radius	5.5 m
Small plasma radius	0.53 m
Magnetic field	3 T
Duration of discharge	Up to 30 min
Plasma heating	14 MW
Plasma volume	30 m^3
Plasma mass	5–30 mg
Plasma composition	Hydrogen, deuterium
Plasma temperature	60–130 million °
Plasma density	3×10^{20} particles/m^3

combustion residue, which as a noble gas is completely harmless to the environment. Depending on the type of plasma inclusion technologies future fusion reactors will have a diameter between 10 and 45 m.

6.1.3 The Stellarator

Contrary to a Tokamak machine, in which pulsed electrical current is conducted directly through a plasma, stellarators generate a magnetic field by electric current flowing through coils outside the plasma vessel. The Wendelstein X-7 of the Max-Planck Institute for Plasma Physics in Greifswald in Germany is the biggest stellarator plant in the world. It consists mainly of 50 superconducting magnet coils, which are projected to enable plasma discharges of 30 min duration at 100 million °C.

The construction was finished in 2014; the first hydrogen plasma with a pulse duration between 0.5 and 6 s was accomplished in 2015. The associated temperatures measured amounted to 100 million °C for electrons and 10 million °C for ions. It followed an upgrade lasting 15 months before the second experimental phase was initialized in 2017. It is hoped that in the near future discharges of a duration of 30 min. can be achieved. Table 6.1 summarizes the most important technical parameters.

6.2 Fuel Cell

As indicated by the term, fuel cell stands for a technique, which enables the combustion of certain substances to achieve a transformation into electrical energy. Fuels can be

- natural gas,
- bio gas,

Fig. 6.2 Fuel cell

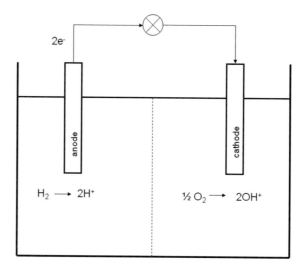

- but mainly hydrogen.

Liquids with a low boiling point such as alcohols can be used as well. The operating principle of a fuel cell is similar to that of a battery (Fig. 6.2). The setup consists of two electrodes, the fuel itself and oxygen together with a semi-permeable dividing membrane.

The combustion reaction is the combination of hydrogen and oxygen to water (H_2O). This reaction releases chemical binding energy.

In a fuel cell, hydrogen and oxygen react with the help of catalysts. The catalysts (noble metals) are housed in the electrodes themselves. The process proceeds as follows:

The catalyst at the anode extracts 4 electrons, thus creating two positively charged hydrogen ions (protons) each from the H_2. These migrate through the semi-permeable membrane into the oxygen region. At the same time, the electrons taken from the hydrogen are lead to the oxygen, which absorbs the electrons via the cathode and thus reacts with the hydrogen ions to create water. The relevant equations are

$$\text{Anode: } 2H_2 \rightarrow 4H^+ + 4e^- \tag{6.6}$$

$$\text{Cathode: } O_2 + 4\,H^+ + 4e^- \rightarrow 2H_2O \tag{6.7}$$

Thus between anode and cathode, an electrical current is flowing. Now an electrical consumer can be placed into this circuit. However, fuel cells manufactured up to now generate only a small voltage (<1 V). Therefore, many cells are connected together as stacks. Over time the fuel and thus the cell itself is used up. But since the fuel can be reloaded in an ongoing process fuel cells continue to function.

Fuel cells do not produce any harmful substances directly, only water. There are a number of different small applications for fuel cells, for example, in mobile phones or hearing aids. The most important field of research concerns automobile power units. This research is pushed by car manufacturers. One important aspect among others concerns the handling of the highly explosive hydrogen. Another research object deals with the design of efficient catalysts.

Chapter 7
Smart Energy

Abstract The energy transition the whole world is talking about is not only a gigantic changeover of energy technologies. Since certain energy objectives cannot be achieved by these means alone, and since increasing demand caused by consumer behavior has to be taken into account, the turn to smart energy has become a spark of hope. This hope is based on a vision, where intelligent control of energy usage may lead to less capacity and at the same time to counter steering an ever increasing consumer demand.

7.1 Introduction

The energy transition the whole world is talking about is not only a gigantic changeover of energy technologies. Since certain energy objectives cannot be achieved by these means alone, and since increasing demand caused by consumer behaviour has to be taken into account, the turn to smart energy has become a spark of hope. This hope is based on a vision, where intelligent control of energy usage may lead to less capacity and at the same time to counter steering an ever-increasing consumer demand.

Such scenarios are basically feasible from a technological point of view. There is no question about it that the required smart systems are prone to unwanted disruptions or targeted attacks from outside. Furthermore, they require the disclosure of private, in part intimate information on a scale hitherto unprecedented—risking all sorts of consequences by abusive usage.

This section will deal with the smart energy vision and its technological prerequisites.

7.2 The Smart Energy Vision

The smart energy vision is determined by three stakeholders:

- the legislator,

© The Author(s), under exclusive license to Springer Nature Switzerland AG 2022
W. Osterhage, *Energy Utilisation: The Opportunities and Limits*,
https://doi.org/10.1007/978-3-030-79404-0_7

- utilities,
- the customer.

but also by different technical possibilities, which are constantly refined. The realization of this vision has to take into account two major components:

- decentralized electricity generation,
- central electricity generation.

Decentralized electricity generation takes place locally, i.e. by the end customer, for example, by photovoltaic assemblies, combined heat and power plants, geothermal heat, small wind power plants or biogas facilities. As a first step, they should ensure the self-supply of households and small enterprises. In the case of surplus electricity, this can be fed into the grid. If the weather is unfavourable or for demands of energy-intensive equipment additional electricity can be drawn from the public grid (this may be generated by other decentralized units or central power plants).

Centralized power plants could be:

- large scale conventional power plants,
- electricity from reservoirs,
- electricity from wind farms,
- electricity from solar power plants, etc.

The intelligent control of a smart energy network is supported by a special system of tariffs by the supplier depending on

- the availability of the energy on offer,
- consumer behaviour.

For example: if a lot of electricity is generated during periods of decreasing demand (at night), favourable tariffs can be offered to the consumer at short notice. For this, two conditions have to be met:

- communication of consumer routines from the consumer to the supplier,
- communication of tariff changes in due course to the consumer.

Certain technical requirements are necessary to realize this.

7.3 Smart Meters

Smart meters replacing traditional ones allow for bidirectional communication between consumer and supplier. Meanwhile, in some countries, they are obligatory for all new buildings. On the one hand, they facilitate—after technical implementation—the transmission of information concerning consumption data from any appliance to the supplier, on the other hand, actual tariffs are received by the consumer in real time.

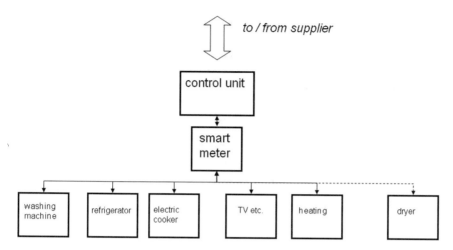

Fig. 7.1 Smart metering

Here are some of the data to be transmitted to a supplier from a smart home:

- operating times and consumption of large household appliances (washing machine, dryer),
- rising hours of family members (energy consumption during showers),
- switch on of small appliances (coffee machine, toaster),
- usage times of computers and electronic entertainment media,
- heating routines etc.

In the end, a network as shown in Fig. 7.1 is a prerequisite for the realization of the smart energy vision.

Once electrical cars are integrated into the energy circuit data about journey time, destinations (electric charging station, booking of consumption to personal credit cards), etc. can be provided.

7.4 Smart Grid

A smart grid, i.e. an intelligent network, does not only connect all sorts of suppliers and consumers, resorting to any of the resources presented in this book, but there are some additional conditions that have to be met in the first place:

- algorithms capable to calculate contour integrals for the matching of supply with demand,
- pricing systems depending on supply and demand (including contract management),
- communication means between consumer and electricity supplier.

By bringing together all these elements something like a virtual power plant with autonomous control can be brought into existence. A virtual power plant combines, for example, classical large-scale power plants in connection with individual small plants from wind power and photovoltaic. This network operates under its own optimization strategy with regard to baseload or peak consumptions.

Smart metering closes the gap between the Internet of Things (smart home) and ubiquitous computing. The synthesis between energy and information management would be achieved. The consumer data collected would allow for establishing exact customer profiles with relevant load curves, which can then be used by the suppliers. This is not the place to discuss the associated problems of data protection and risks. They can be found in the relevant IT specialist literature.

Chapter 8
Appendix: Climate Engineering

Abstract There is a possible connection between energy technologies and climate change. The set point and basis for the following discussion are: there is climate change and there are considerations to deliberately and artificially influence it. In the context of pro-active interventions in the climate system there are two fundamentally different technological approaches, which will be explained in detail:

- causal removal technologies and
- technologies for symptomatic compensation of climate change.

What will be the consequences of such acts? What are the possibilities and associated risks? How do physical processes look like in this context? What are the ethical aspects touched upon?

8.1 Point of Departure

There is a possible connection between energy technologies and climate change. However, at this point, we do not want to start off a fundamental debate about climate change. This is not the object of this section. We take into consideration:

- Short-term observations (within the framework of some decades, at least since the onset of systematic data acquisition with acceptable quality) lead to the conclusion that the planetary climate changes.
- Comparison with coarse data from the past, dating further back, support this trend analysis.
- Based on these developments, extensive model calculations have been undertaken to try to project further climatic development into the future.

Against this backdrop, the conclusion is that there is climate change, which could be disadvantageous at least for parts of if not for all of the planet. There have always been phases of climate change during geological history, sometimes with dramatic consequences for nature. Smaller and locally confined climate variations have happened while human civilization already existed. Now, one can try—as is done already—to differentiate theoretically between natural cycles and the consequences of human endeavors. The set point and basis for the following discussion are: there is climate change and there are considerations to deliberately and artificially influence it. What will be the consequences of such acts? What are the possibilities and associated risks? How do physical processes look like in this context? What are the ethical aspects touched upon?

8.2   The Basic Problem

The simple question to be answered is: what can be done in face of the backdrop described above? There are three possibilities:

1.   To do nothing, wait and see, if the problem goes away by itself, and hope that there will be some advantages for one part of the world populace (from a geographic point of view) compensating for disadvantages suffered by another part.
2.   To continue with measures for the reduction of greenhouse gases, for example, wherever appropriate accelerate these measures, hoping that they will suffice, in as much as the strengthening of them can be kept up politically.
3.   Initialize proactive interventions in climate dynamics; this is the subject of our considerations in this section.

All three measures, especially the latter, have to be subject to risk analysis in view of the successes expected. This will be tried qualitatively with respect to point 3.

Figure 8.1 gives a schematic overview of options to intervene in the climate system.

8.3   Selective Interventions in the Climate System

In the context of proactive interventions in the climate system, there are two fundamentally different technological approaches, which will be explained in detail:

- causal removal technologies and
- technologies for symptomatic compensation of climate change.

How is the state of public debate about this problem today? What consequences have to be expected, and what are the possibilities concerning their predictability? Are there noteworthy field experiments, and what impact do they have? It is to be

Climate Engineering

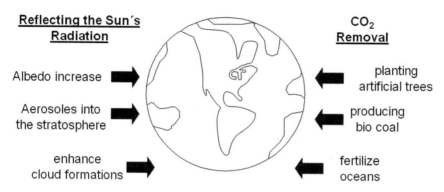

Reflecting the Sun's Radiation

Albedo increase →

Aerosoles into the stratosphere →

enhance cloud formations →

CO₂ Removal

← planting artificial trees

← producing bio coal

← fertilize oceans

Fig. 8.1 CE options

assumed that—as with any major technology—CE will harbor conflict potentials up to a danger of international conflicts. Independent from technical feasibilities financial costs for realization will certainly play a role permitting realization or not.

8.3.1   Causal Removal Technologies

Just as technologies for compensation, these measures will be discussed in detail in the following chapter. At this stage, we will only give a short overview. The main approach is the so-called CDR approach. CDR means "Carbon Dioxide Removal", i.e. the elimination or removal of carbon dioxide already produced. To reach this objective different

- biological,
- chemical,
- physical processes,

have to be tried to let CO_2 be absorbed by oceans and the biosphere.

8.3.2   Technologies for Symptomatic Compensation of Climate Change

Technologies for compensation can be grouped under the term "Radiation Management" (RM). This refers to attempts to influence the sun's radiation onto the earth. This could be achieved by three different approaches:

- reduction of short wave sunlight,
- increase of atmospheric reflection or reflection from the earth's surface,
- increase of longwave thermal emission back into space.

8.3.3 State of Discussion

Public debate but also discussions among specialists are still at an early stage. Wide sections of the public are not familiar with the subject at all. Occasionally pertinent media do publish reports. The actual debate takes place in a small circle of experts. Up to now, the stakeholders come from:

- research,
- some NGOs,
- interested enterprises,
- politics.

On the one hand, research deals with general considerations about radiation balance, on the other hand with the development of dedicated technologies. This means that expected impacts, their efficiencies with respect to the objectives cited and resulting consequences for society regarding acceptance problems during the preparation phase or by climatic changes after the deployment are assessed completely differently by the involved persons.

In most countries, there is widespread non-transparency with respect to planning and objectives. Together with the associated uncertainty not yet articulated audibly this holds the possibility of conflicts and potential for polarization. It is likely that an overall consensus in society will not be achieved. In any case, a complex debate can be expected. Although we are at an early stage yet there are already advocates and adversaries.

Adversaries are worried that acceptance of CE will among other things lead to relaxing efforts for emission control. Therefore, some people think CE can be made superfluous by intensifying emission control. Furthermore, the efficiency of CE measures including their economic impact is questioned. It is feared that there will be unwanted side effects (we will discuss this argument further down). And last but not least ethical aspects play an important role in CE refusal.

Advocates argue that CE will be more effective than emission control in any case. Furthermore, they put forward that climate objectives defined by the world could never be reached without CE. In any case, CE should be kept as an emergency option, if otherwise a climate catastrophe cannot be avoided (but since some CE approaches need long time spans to become efficient, this argument does not appear to be plausible).

8.3.3.1 Consequences and Predictability

The first question to be asked is: what shall be compensated in any case? For this one has to look at both technological approaches in detail. In theory, RM technologies do render a quick decrease in global temperature but are less efficient to manage rainfall distribution for example. Furthermore, they have to remain in place for reasons of sustainability.

If the objective is to roll back already experienced climate change to a state still to be defined, this is only possible by CDR technologies. However, they do not promise a quick reduction in average global temperature.

When building an overall picture including all possible scenarios, basically all material and energy flows have to be taken into account. These cycles by their nature react the more sensitive the higher the impact of technology deployment is.

RM technologies intervene in the global radiation balance. What is still completely unknown is its feedback to remaining earth systems as well as a possible impact on the biosphere. CDR technologies may influence biological cycles as well. Furthermore, meteorological side effects caused by them are still not predictable. But should it not be possible to get rid of such uncertainties by research efforts—if necessary at a large scale—before its effective deployment? To this end, the following can be said:

The earth's system is sufficiently complex that insights gained at a regional level cannot lead to specific conclusions about the real global impact and side effects to be expected. Therefore for this reason alone, a risk-free CE is unthinkable. This means we would have to do with a further anthropogenic factor when tailoring the climate. Of course, people think about large-scale field experiments. But such experiments need—depending on the technology to be deployed—long periods of observation, sometimes several decades. The important monitoring associated with these has to be capable to distinguish between natural and artificial long-term effects. This means that natural climate cycles have to be understood much better, which is still not the case. In any case, the societies in the participating countries would have to bear an enormous burden. The associated discussion would be comparable with those regarding nuclear energy or genetic engineering.

8.3.4 Legal Context

Although this is a technical-scientific book, I would still briefly like to touch upon the legal context. The measures considered so far are transfrontier approaches. They could be launched by individual states, but would have direct consequences for other states. This immediately implies the application of international law. But for this, there does not exist any legal reference base. Neither have binding norms been developed nor does a generally accepted definition of CE exist. This means that comprehensive agreements have to be drafted in the first place.

8.3.4.1 Conflict Potential

Considerations of CE had hardly begun, when deliberations already drifted into a completely different direction—the usual dual-use potential: can CE technology also be deployed militarily? The idea of a climate weapon was born. The military in some countries has already undertaken feasibility studies. They have come to the following conclusions:

The deployment of a climate weapon is not entirely unlikely, especially when one believes to be able to generate locally confined weather modifications on the battlefield, which would be disadvantageous to the enemy. At the same time, military usefulness is classified as negligible. Furthermore, regional confinement is, as already mentioned, difficult if not impossible. Since the deployment of a climate weapon would constitute the breach of international law the ensuing political costs would be high, so that really only non-state actors would potentially use it. But since today most international conflicts are determined by the latter, such a danger should not be underestimated.

8.3.4.2 Institutional Integration

International cooperation concerning CE is unavoidable taking into account, what was outlined so far. This comprises coordination of research projects (if only because of costs), but also an independent control body. The first task would be the approval of binding guidelines. Another useful project would be to compare current and future agreed-upon measures and results of emission control to the deployment of CE. And finally, exit modalities would have to be clarified.

8.3.5 Costs

Today the state of knowledge regarding costs is rudimentary. Practically, only estimates of possible operating costs for the different technologies under consideration are provided. Of course, expenditures for conducted or current research projects are known, however, there are no reliable propositions concerning the large-scale research projects necessary, if it is decided to follow the CE path. There are especially no reference points for follow-up costs from side effects or for compensating measures. It is certain that CE measures will influence the economic systems of specific regions and countries. Overall economic effects over long time spans are not foreseeable.

8.3.6 Approaches

The question is still: emission control or CE or both? Already now opinions are drifting apart. Some experts fear that with CE emission control will subside. Again costs would be one of the arguments. In this case, it is assumed without sufficient evidence that emission control is more expensive than CE measures. But because of the uncertainty regarding costs as outlined above this remains pure speculation.

One way out of the cost trap would be the answer to the question: could CE be driven commercially? In this case, cost risks would partly be taken out of the budgets of participating countries and transferred to the private sector. However, in a commercial context, the danger exists that it may develop its own dynamic driven in turn by commercial criteria. Therefore, regulative standards and competition control would be required besides national incentives.

Altogether the direction points to an integrated approach. Emission control would be kept besides Climate Engineering. Then other human factors would have to be included, like, for example:

- land utilization,
- agriculture,
- surface modifications.

An intensive academic discourse between potential research institutions has to be initialized. Such a discourse should be spread over all of society and include the dimensions

- social,
- ecological,
- economical,

to form the basis of an overall climate-political concept.

8.4 Detailed Technological Measures

8.4.1 Starting Point

The energy balance of the earth's system is determined by two factors:

- the solar constant S (1370 W/m^2): incoming short wave radiation,
- albedo A: the total reflection of the sun's radiation by the earth's system (30%).

Thus we can differentiate

- short-wave radiation flux: $F_k = S(1\text{-}A)$,
- and long-wave radiation flux emitted from the earth: F_l.

In case of $F_k = F_l$, there is equilibrium, i.e. the amount of incoming radiation energy is completely reflected, so that no residual warming can be expected because of natural causes.

8.4.2   Possibilities of Influence

How can the radiation balance be influenced? There exist basically three possibilities:

- decrease of the solar constant or
- increase of albedo
- increase of thermal emissions of the earth system F_l.

This leads to two types of methods:

- Solar Radiation Management (SRM),
- Thermal Radiation Management (TMR).

Table 8.1 gives an overview of possible CE measures in this context.

8.4.2.1   Reduction of Irradiation

To reduce the irradiation onto earth the following approaches are discussed:

- Placement of millions of mirrors at the Lagrange point between sun and earth (1.6 mio. km from the earth) for a period of several decades. The Lagrange point is the position in our solar system, where the attractions of earth and sun are neutralized (Fig. 8.2). This measure would lead to a reflective cloud of 100,000 times 13,000 km^2.
- Placement of reflective umbrellas in an orbit around the earth (Fig. 8.3).
- Creation of a pulverulent layer of dust with a total mass corresponding to a medium asteroid near the earth (Fig. 8.4).

Table 8.1 Overview of CE measures

SRM	Decreasing incoming radiation	Measures in space
	Increasing the reflection of the sun's radiation	Modification of the Earth's surface
		Manipulation of marine sheet clouds
		Manipulation of the stratosphere
TRM	Increasing emission	Manipulation of cirrus
	CDR measures	Physical
		Chemical
		Biological

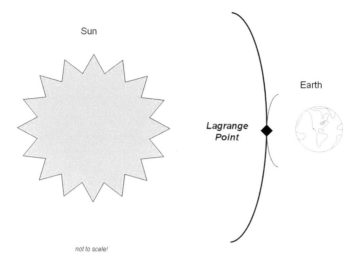

Fig. 8.2 Lagrange point

Fig. 8.3 Umbrellas in earth orbit

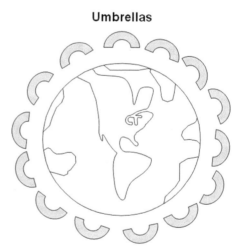

There exist model calculations for all these measures with results pointing to different effects in all directions, depending on the choice of various parameters available.

Fig. 8.4 Dust in earth orbit

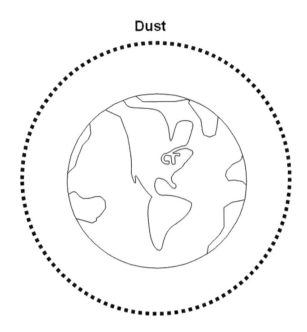

8.4.2.2   Increase of the Reflection of the Solar Radiation

Alternatively (or in combination?), there are approaches to increase the albedo. This would be achieved by selective surface modifications. Table 8.2 shows some common albedo values.

Increasing albedo can be achieved by a variety of measures.

Increase of albedo by reflecting dust particles

One approach would be to insert sulphur dioxide into the stratosphere. The salt particles would serve as concentration nuclei for the creation of water droplets in the clouds. There are serious arguments regarding consequences other than changing the albedo. One concerns the possible generation of sulphuric acid droplets. This could lead to possible warming of the stratosphere and to an increase of acidity of rainfall with negative effects on the global water cycle.

Table 8.2 Albedo values in [%]

Surface	Albedo
Snow	80–90
Clouds	60–90
Desert	30
Arable land	26
Forest	5–18
Water	3–22

Modification of Marine Sheet Clouds

Clouds, which hover close above the ocean surface, are called marine sheet clouds. To achieve the planned objective, ships would be deployed to collect seawater by pumps and then spray this water with special siphons into these low hovering clouds. In the end, this would lead to an increase in the cloud albedo.

Increase of Albedo by Reflecting Land Surfaces

To increase the albedo of reflecting land surfaces there are two possible methods:

- modification of agriculturally cultivated land,
- intentional modification of urban areas.

While the latter option can be realized relatively easily (apart from the costs and consequences for the inhabitants of towns), the first approach is more difficult. To increase the albedo, the cultivation of plants with a high leave gloss is necessary. This is only possible by enlarging unfertile land areas, since this effect cannot be achieved with traditional crops. The resulting consequences would affect biodiversity and influence the carbon cycle in an incalculable way.

8.4.2.3 Increase in Thermal Emissions

The next important complex of methods deals with increasing thermal emissions. The following alternatives are on offer:

- modification this time of cirrus,
- acceleration of the physical carbon pump in the oceans,
- acceleration of the physico-chemical carbon pump in the oceans,
- acceleration of the biological carbon pump in the oceans,
- increase in carbon sequestration,
- acceleration of weathering,
- controlled removal of CO_2.

Modification of Cirrus

This time airplanes would be deployed for this purpose. They should be used to seed ice crystals into the cold high ice clouds. This would lead to a modification of emission effects. Today there are still significant uncertainties regarding the dynamics of ice nuclei in ice clouds. Frequent repetition of seeding would be necessary since cirrus are inherently unstable and do not remain in the same place.

Physical Carbon Pump

What is meant by this?—Oceans are known to be CO_2 sinks. It is planned to modify descending ocean currents by different technologies in such a way that the absorption of CO_2 in the deep sea is intensified. Concerns about these measures will be discussed further in Sect. 8.4.4.

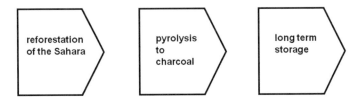

Fig. 8.5 Pyrolysis

Physico-chemical Carbon Pump

The physico-chemical carbon pump uses the dissolubility of CO_2 in water. This could be achieved by introducing lime minerals into the seas to increase their alkalinity and thus strengthen the dissolubility (comments under Sect. 8.4.4).

Biological Carbon Pump

This approach intervenes directly into the food chain. In the first place, it is planned to increase bacteria and algae masses, which would directly lead to an increase in global photosynthesis. A further component is the introduction of iron and nitrogen to increase the growth of plankton. Both measures would modify the natural food chain. It cannot be excluded that the quantity of toxic microorganisms will increase at the same time.

Carbon Binding

Basically, this is nothing else but the burning of wood. This is shown in Fig. 8.5 schematically.

First of all, the reforestation of the Sahara is planned. Once this has been achieved the wood harvested could be treated to become charcoal to bind the CO_2. Then products of this pyrolysis process must be transferred to a long-time store and kept there.

Weathering

It is planned to accelerate the natural weathering process. This idea rests on the perception that weathering by carbonic acid plays an important role as a CO_2 sink on a geological time scale. How could such an acceleration of this natural process be achieved?—Silicate weathering, which is meant here, could be increased by introducing Olivin powder into tropical rain forests or into coastal waters. A first estimate of the quantity required amounts to the equivalent of the present global coal production. Another possibility would be the introduction of hydrochloric acid on land. It is speculated to obtain the required quantity of hydrochloric acid by electro-chemical withdrawal from the sea.

Removal of CO_2

The necessary steps are shown schematically in Fig. 8.6.

Fig. 8.6 Air capture method

The so-called Air Capture Method starts by passing air across an absorber (in this case sodium hydroxide) with pure CO_2 left as a residue. This CO_2 then has again to be transferred to final storage and kept there.

8.4.3 State of the Art

What does the state of the art concerning the measures described under Sect. 8.4.2 look like today? To summarize: most of the presented proposals are based on theoretical papers having been published and continue to be published in ample variety. A multitude of patents has been registered. There are also lots of model calculations, but they all come to different results. Most of them are pure estimates based on non-robust assumptions.

The main problem is that meaningful field tests are only possible on a global scale. Laboratory experiments are only useful to evaluate a specific method itself. Methods to bind carbon are understood and proven for example. The chemical processes concerning weathering are of course known as well. Otherwise, the state of affairs is rather flimsy.

8.4.4 Side Effects

In view of the fact that some of the proposals seem to be partly adventurous the question arises about possible side effects when deploying such technologies. These side effects would of course be different with respect to the type of measures. The following catalogue is only an outline of possibilities without going into the details:

- cooling of the tropics (albedo measures),
- warming of seawater (combination of different approaches),
- general weakening of water cycles,
- modification of the overall energy balance of the earth (albedo and emission measures),
- increase in acidity (acceleration of weathering),
- influence on the carbon cycle (CO_2 measures),

- acidification of the seas (combination of different approaches),
- influencing the food chain (practically by all measures),
- interference in agriculture and fishing,
- increase in alkalinity (carbon pump).

The combination of different CE measures can lead to compensations, opposite effects or over modulation. For large-scale implementation, the whole problem of control and counter steering has to be solved in advance.

8.5   Physical Background Considerations

Next, we want to take a look at some scientific aspects to be observed when implementing CE ideas. Thermodynamic considerations play an important role in this context.

8.5.1   Energy Balances

For this, we refer to the physical basics in Chap. 2.

8.5.2   Chaos

Chaos theory gives additional clues concerning system behaviour. Chaos theory is called a branch of non-linear dynamics. It deals with orders in dynamic systems, the development of which over time seems to be unpredictable.

In chaotic behaviour, even extremely small alterations of the initial values lead to completely different properties after some time. Chaotic behaviour happens in systems, the dynamics of which is described by non-linear equations. Reasons for exponential growth with respect to differences in the initial conditions are often mechanisms of self-energizing by back coupling for example.

Most natural processes are based on non-linear behaviour.

The reliability of weather forecasts, for example, is limited by the rudimentary knowledge of initial states. But even based on thorough information, long-term weather forecasts would fail in face of the chaotic nature of meteorological events.

Figure 8.7 shows schematically the development of a sand heap, to which grains of sand are fed continuously—as long until it loses its stability and its shape starts

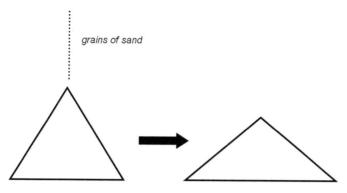

grains of sand

Fig. 8.7 Quasi-chaotic sand heap

to slip—triggered by a final grain of sand, which causes the inherent quasi-chaotic state to tilt.

8.5.3 Fine Adjustment

Before consolidating our scientific considerations so far, I would like to draw attention to some facts regarding the fine adjustment of our planet and thus life on earth:

There are many areas of life, which are constantly in a quasi-chaotic state:

- the stability of economic systems,
- physical systems like stars or liquids and gases.

Concerning the latter, one example is the mass difference between neutrons and protons. If this would only be slightly different from what it actually is there would be no nuclear physics in the conventional sense, no elements as well and no stars. Or, if the energy level in the ^{12}C nucleus would be different from 7.65 meV there would be no life based on carbon chemistry.

It looks as if many constants in nature are within such limited boundaries that make human life possible. These constants cannot be influenced by us. But there are further aspects of fine adjustment within our fragile area of life. They concern:

- the balance between landmass and oceans
- the composition of the atmosphere,
- the distribution of the diversity of animals and plants across our planet.

These and some other factors would be affected by CE.

8.6 Consolidation

Let us now consolidate the insights from Sect. 8.5; we realize:
  Climate-Engineering measures

- are irreversible processes,
- are capable to disturb the sensitive fine adjustment of the area of our life (in a quasi-chaotic state), can develop a chaotic dynamic.

8.6.1 Irreversibility

Claim: *A controlled exit from CE scenarios will not leave significant damage.*

Counterclaim: Any intervention into the earth system is irreversible. In certain cases, this may lead to irreparable catastrophes, which are more damaging than climate change up to now.

> In the end the final result would be the opposite that CE wants to prevent: the endangerment of life on our planet.

8.6.2 Arguments

Further advocating arguments can be put forward against the above concerns:

- There have always existed similar interventions.
- The construction of houses and towns has changed the surface of the earth sustainably.
- Large-scale clearing of woods has led to the creation of deserts.
- Emission of exhaust gases has changed the atmospheric composition.
- In the course of the development of our civilization, we continuously intervene in the environment.

  For all these reasons, CE would not only be legitimate but even demanded.
  There is a simple argument against this: we do not possess a suitable test system to check every consequence before going live as is standard procedure in the case of introducing new software systems for example. Figure 8.8 illustrates that we have only one shot! If this fails there is no way back.

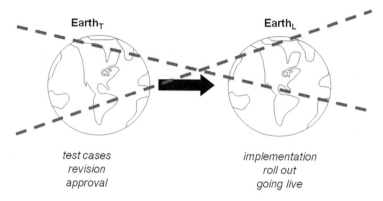

There is no test system!

$Earth_T$ $Earth_L$

test cases   implementation
revision   roll out
approval   going live

Fig. 8.8 One shot

8.6.3 Risk Analysis

Since we do not possess sufficient data statistical forecasts on the basis of for example normal distribution algorithms relative to risks are not possible. Other considerations have to be taken into account, for example, those of Nassim Taleb, who has developed the Black Swan Theory. He demonstrated that conventional statistical investigations make only sense in very controlled environments. On the other hand situations in all areas of life may arise, which are suddenly found to be not only outside deterministic scenarios, but may be based on data, which cannot be predicted even by classical statistics. For such scenarios, he proposes a differentiated approach by assuming that there always exists one more risk, which has not been taken into account (Fig. 8.9). Taleb calls such unprobable events a "black swan". Why? Until the discovery of Australia, all Europeans were convinced that only white swans existed. This was based on observations lasting thousands of years. But then the black swans emerged in Australia: against all probabilities.

8.7 Reference Frame

Supposed in the end it is decided to launch CE, what will be the objectives? The earth has passed through many periods of climate variations and environments hostile to life. The aim is certainly not to relapse into such times. But where do we want to go by CE? There may be the following reference points for example:

- to keep the status of today,
- to go back to the status of 50 years ago or even,
- to reset to the status of before 100 years?

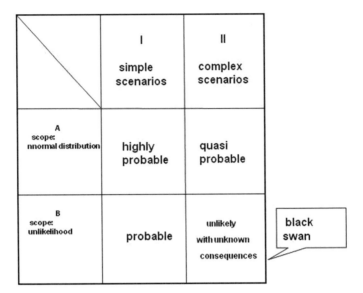

Fig. 8.9 The black swan

The decision has to be taken, whether CE should only be used to correct some side effects of industrialization. If that were the case the state before 1750 would be relevant (at that time it was rather cold: the aftermath of the last minor ice age).

In this context it has also to be decided, who would have the authority to determine such objectives. Which is the eon to base the guidelines upon? At the same time, one has to bear in mind that the earth today is not habitable everywhere (deserts, the poles) and this already since before industrialization. All these considerations may lead to the desire by certain countries to enhance their present habitat by applying CE. Or is it that in the course of CE the overall living environment of the earth should be optimized?

8.8 Ethics

There remains a final dilemma independent of all technical considerations:

If we could indeed achieve all this, are we free to do so at all? There are some basic considerations behind this:

- Does economic benefit outweigh the gift of creation?
- Have we neglected our garden?
- Do we want to play God?

In spite of all statistical considerations regarding cosmological estimates about the number of possible life-supporting environments in the universe, it is still true that.

- the earth is a unique planet at a lonely place in the universe,
- the possibility of our life needed a unique fine adjustment,
- we have a singular unique responsibility to maintain this precarious life-supporting domain, which always resides in an unstable equilibrium.

This we, as rational creatures, should accept as our mission.

8.9 Conclusions

To summarize: CE approaches exist today as theoretical possibilities. There are no practical experiences with them up to now. These could be gathered by using suitable test systems. However, one has to bear in mind that even test systems can lead to irreversible consequences. But if all reductions alone do not suffice could CE under careful risk analysis be the final resort? Is it possible to maintain CE in a kind of standby mode? Would there be time enough for it to be effective?

The decision between possible CE consequences or idleness demands a high political responsibility concerning future global development.

Problems and Solutions

Problem 1: Compilation of an Exhaustive Energy Balance

- production of raw materials, processing, disposal of waste, all intermediate transports for

  – fuels,
  – materials for component construction,
  – operating supplies.

- construction of components including

  – structural requirements (factories),
  – production materials (machines),
  – disposal of wastes,
  – all intermediate transports.

- plant construction ....
- operating supply infrastructure ....
- disposal of ashes ....
- ....

Taking economic considerations into account quite different requirements have to be added:

- infrastructure like roads for transports,
- railways and industrial premises,
- regeneration of labour,

and all overheads for administration and management, which have to be respected for an honest balance.

© The Editor(s) (if applicable) and The Author(s), under exclusive license
to Springer Nature Switzerland AG 2022
W. Osterhage, *Energy Utilisation: The Opportunities and Limits*,
https://doi.org/10.1007/978-3-030-79404-0

Problem 2: Calculation of Specific Weight

weight of a body in air: 27 g,
weight of the same body in water: 16.5 g,
reduction in weight 27 g − 16.5 g: 10.5 g,
volume of the body: 10.5 cm³

Hence, specific weight: $\rho = m/V = 27/10.5 = 2.57$ g/m³.

Problem 3: Register Tons

There is no relationship between register tons and the Archimedean Principle. Register ton is an ancient cubic measure, which has to do with displacement only indirectly and is used in shipping.

$$1 \text{ RT} \mathrel{\hat{=}} 100 \text{ feet}^3 \approx 2.8 \text{ m}^3$$

Gross register ton: comprises the whole vessel,
Net register ton: difference between gross register ton and common areas like lodgings, engine rooms, etc

Problem 4: Calculation of Photon Energy

$$E = h\nu$$

$$c = \lambda\nu$$

$$E = hc/\lambda = 7.95 \times 10^{-19} \text{J} = 4.96 \text{ eV}$$

Problem 5: Calculation of Kinetic Wind Energy

$$\Delta m = \rho \Delta V = \rho \, Av \, \Delta t$$

with ΔV a volume element and ρ the specific weight. The kinetic energy is then calculated as

$$E_{kin} = \Delta mv^2/2 = \rho \, Av^3 \Delta t/2.$$

Problem 6: Calculate the Annual Equivalent of Fuel Oil Produced by a Cow

The average calorific value of bio gas amounts to 4–7.5 kWh/m³. Thus the average calorific value of one cubic meter bio gas corresponds to about 0.6 l fuel oil.

This means a single cow produces the equivalent of about 300 l of fuel oil per year as manure.

Index

A
Absolute temperature, 12
Absorbtance, 60
Absorption, 77
Accumulator, 111
Acidity, 161
Activity, 96
Adiabatic exponent, 53
Adiabatic system, 14
Air capture method, 161
Air masses, 112
Albedo, 155
Alcalinity, 162
Alpha-rays, 91
Alternating current, 36, 114
Amorphous silicon, 111
Anaerobic, 121
Anergy, 21, 23
Angular frequency, 41
Angular velocity, 79
Annihilation radiation, 95
Anti-neutrino, 93
Atmospheric composition, 164
Atomic mass, 32
Atomic nucleus, 83
Atomic number, 75, 83
Atomic particle, 74
Atomic weight, 82
Average life time, 97
Axial turbine, 54

B
Backwater effect, 131
Balmer Series, 77
Band spectrum, 77

Battery, 37, 143
Bernouilli's Equation, 50
Beryllium, 108
Beta-rays, 91
Big Bang, 5
Binding energy, 67, 84
Biological cycle, 153
Biosphere, 151
Blackbody, 61
Black Swan Theory, 165
Blading, 54
Borehole, 125
Boron, 101
Boyl's Law, 12
Buoyancy, 48

C
Cadmium, 101
Cadmium telluride, 111
Capacitor, 34
Capacity, 34
Carbon, 101
Carbon chemistry, 163
Carbon cycle, 159
Carbon dioxide, 119
Carbon dioxide removal, 151
Carbonic acid, 160
Carbon monoxide, 119
Catalyst, 143
Causal removal technology, 150
Celsius, 12
Chain reaction, 100, 107
Chaos theory, 162
Charging station, 147
Chip, 116

Climate change, 149
Climate dynamics, 150
Climate system, 150
Climate weapon, 154
Combustion engine, 103
Combustion gas, 105
Combustion reaction, 143
Combustion technology, 117
Compressible gas, 53
Compton Wavelength, 73
Concentrator, 109
Conservation of energy, 14
Consumer, 146
Coriolis force, 113
Coulomb attraction, 79
Coulomb repulsion, 84
Coulomb's Law, 32
Critical mass, 100
Cut-off frequency, 67

D
Dark Energy, 7
Dark Matter, 7
Data protection, 148
Decay, 94
Decay constant, 96
Deep thermal energy, 125
Dehydration, 118
DEMO, 140
Density, 47
Desulfurization, 124
Deuterium, 141
Dipole, 38
Distribution network, 133
Downpipe, 129
Down-quark, 93
Dry wood, 116
Dynamic viscosity, 47
Dynamo, 40, 129

E
Effective output, 103
Efficiency, 16, 69
Elastic scattering, 86
Electrical car, 147
Electrical circuit, 34
Electrical current, 33
Electrical tension, 33
Electric dipole, 75
Electric field, 32
Electricity generation, 36

Electric power, 44
Electro-chemical storage, 135
Electrolysis, 36
Electrolyte, 37
Electromagnetic radiation, 57
Electromagnetic wave, 62
Electron, 32, 75
Electron capture, 95
Electron orbit, 75
Emergency aggregate, 106
Emission, 77
Emission control, 155
Emissivity, 61
Energy balance, 23
Energy carrier, 23, 116
Energy level, 79, 84
Energy transformation, 21
Enthalpy, 28
Entropy, 17
Environmental temperature, 25
Equation of state, 53
Euler's Equations, 53
Exchange of energy, 58
Excitation, 76
Exergy, 20, 23, 104
Exhaust gas, 164
External photo effect, 67
Extra-high voltage, 134

F
Fahrenheit, 12
Fast breeder, 108
Fast fission factor, 107
Fermenter, 123
Fission, 85, 99, 107
Fission cross section, 107
Flow law, 47
Flow line, 49
Fossil energy, 105
Fossil fuel, 120
Fresnel lense, 109
Friction, 8
Fruit, 116
Fusion, 85, 99
Fusion reactor, 141

G
Gamma-radiation, 91
Garden waste, 116
Gas combustion, 118
Gas pressure, 31

Gas-steam-combination, 105
Gas turbine, 105
Generator, 40
Geothermal probe, 125
Geothermal reservoir, 125
Global temperature, 153
Global water cycle, 158
Gluon, 93
Graphite, 135
Gravitation, 32
Greenhouse gas, 150
Ground collector, 125
Ground state, 76

H
Half-life, 92, 97
Heat, 15, 23
Heat exchange, 18
Heat flow, 18, 19
Heat pump, 124
Heavy water, 108
Helium, 92, 108, 141
High temperature reactor, 108
High voltage, 134
Hydrocarbon, 119
Hydrochloric acid, 160
Hydrogen, 77, 144
Hydrogen bomb, 139
Hydrogen sulfide, 122
Hydrostatic pressure, 48

I
Impossible process, 16
Intermediate storage, 133
Internal energy, 13, 107
Internal photo effect, 67
Internet of Things, 148
Inverter, 111, 114
Ion, 76
Ionization energy, 76
Irreversible process, 16, 164
Isobar, 26
Isobreeder, 109
Isotherm, 11, 18
Isotope, 82
ITER, 139

J
Junction, 69

K
Kaplan turbine, 126
Kelvin, 12
Kinetic energy, 9, 112, 127
Kirchhoff's Law, 35
Kirchhoff's Radiation Law, 61

L
Lagrange point, 156
Left Hand Rule, 39
Line spectrum, 77
Liquid drop model, 84
Liquid lead, 109
Liquid particle, 49
Liquid sodium, 109
Lithium fluorite, 108
Lithium ion, 135
Lithium peroxide, 136
Loading system, 117
Long wave radiation flux, 155
Low voltage, 134
Luminous flux, 62
Luminous intensity, 62

M
Mach's Angle, 54
Magnetic field, 38
Magnetic flux, 45
Marie Curie, 87
Mass defect, 85
Mass number, 83
Matter, 7
Maxwell's Equations, 42
Mean life time, 99
Mechanical heat equivalent, 13
Medium voltage, 134
Methane, 121
Moment of inertia, 79
Monocrystalline silicon, 111
MOX, 109
Multiplication factor, 107

N
Natural gas, 142
Neutrino, 92
Neutron, 82, 83, 93
Neutron flux, 140
Neutron star, 80
Newtonian liquid, 47
Noble metal, 143
Non-linear dynamics, 162

Nuclear energy, 107
Nuclear reactions, 99
Nuclear reactor, 107
Nucleon, 83
Nucleus, 75

O

Offshore plant, 115
Olivin, 160
Optical model, 84
Orbital angular momentum, 79
Organic matter, 106
Organic solar cell, 111

P

Partial activity, 96
Path line, 49
Pauli Principle, 80
Pellet, 120
Periodic table, 82
Perpetuum mobile, 16
Phase, 18
Photo effect, 110, 111
Photometric law, 63
Photon, 62, 65
Photo synthesis, 120, 160
Photovoltaics, 68
Planck's quantum of action, 62
Plankton, 160
Plasma, 140
Pole, 38
Polonium, 88
Polycrystalline silicon, 111
Positron, 95
Potential energy, 9, 31
Power, 9
Pressure, 11, 46
Probability of decay, 92
Proton, 32, 83, 93
Pyrolysis, 118

Q

Quantization, 78
Quantum hypothesis, 66
Quantum mechanics, 76
Quantum number, 79
Quantum of angular momentum, 83
Quantum orbit, 75
Quark, 93
Quasi-chaotic state, 163

R

Radial turbine, 55
Radiant energy, 59
Radiant intensity, 59
Radiant power, 58
Radiation, 7
Radiation density, 59
Radiation field, 60
Radiation flux, 60
Radiation management, 151
Radioactive decay, 126
Radium, 87
Rain belt, 128
Rainfall distribution, 153
Reactor, 140
Reaumur, 12
Rectifier, 129
Reflectance, 61
Reflection, 60
Reforestation, 160
Resistance, 34
Resonance escape probability, 107
Reversible process, 16
Right Hand Rule, 40
Robert Mayer, 13
Rotor, 42
Rotor blade, 54
Rydberg Constant, 78

S

Scintillator, 7
Semi-conductor, 69
Sensor, 68
Shaft, 54
Shear angle, 47
Shear stress, 46
Shell model, 84
Short wave radiation flux, 155
Silicon cell, 69
Smart home, 147
Sodic aluminum oxide, 136
Sodium, 136
Solar cell, 69, 111
Solar constant, 60, 155
Solar radiation, 109
Solar radiation management, 156
Spectroscopy, 77
Speed of sound, 54
Spin, 79
Stator, 42
Steam turbine, 54
Stellarator, 140, 142
Stratosphere, 158

Strong interaction, 82
Substrate, 69
Sulfur dioxide, 158
Sulfuric acid, 135
Sun belt, 109
Supplier, 146
Symptomatic compensation, 150

T
Technical work, 27
Temperature, 8, 10, 18
Temperature gradient, 18
Tension, 46
Theory of Relativity, 62
Thermal emission, 152
Thermal equilibrium, 10, 12
Thermal radiation management, 156
Thorium, 88
Thorium fluoride, 109
Three-phase alternator, 42
Tidal difference, 131
Tide, 130
TOKAMAK, 140
Total angular momentum, 80
Total life time, 97
Transformation, 103
Transmission, 60
Transmission grid, 133
Tritium, 141

U
Ubiquitous computing, 148
Universe, 5
Up-quark, 93
Uranium, 87, 89, 94

V
Vegetable, 116
Virtual power plant, 148
Voltage, 39
Volume, 11

W
Waste heat, 23, 120, 130
Weak interaction, 93
Weizsaecker equation, 85
WIMP, 7
Wind density, 114
Wind farm, 146
Work, 8, 19
Working substance, 103
Work of emission, 65

Z
Zeroth law of thermodynamics, 11

Printed in the United States
by Baker & Taylor Publisher Services